PE EXAM PREP

CHEMICAL ENGINEERING
SAMPLE EXAMS

Rajaram K. Prabhudesai, PhD, PE Chem. Eng.

This publication is designed to provide accurate and authoritative information in regard to the subject matter covered. It is sold with the understanding that the publisher is not engaged in rendering legal, accounting, or other professional service. If legal advice or other expert assistance is required, the services of a competent professional person should be sought.

President: Roy Lipner
Publisher: Evan M. Butterfield
Senior Development Editor: Laurie McGuire
Managing Editor, Production: Daniel Frey
Quality Assurance Editor: David Shaw
Creative Director: Lucy Jenkins

Copyright 2004 by Kaplan® AEC Education, a division of Dearborn Financial Publishing, Inc.®

Published by Kaplan® AEC Education,
a division of Dearborn Financial Publishing, Inc.®,
a Kaplan Professional Company®

30 South Wacker Drive
Chicago, IL 60606-7481
(312) 836-4400
www.engineeringpress.com

All rights reserved. The text of this publication, or any part thereof, may not be reproduced in any manner whatsoever without written permission in writing from the publisher.

Printed in the United States of America.

04　05　06　10　9　8　7　6　5　4　3　2　1

Please note: Some Kaplan® AEC Engineering titles referenced in this book have changed from their previous editions. The former *Engineer in Training (EIT)* titles are now in the *FE Exam Preparation* series.

Contents

Preface	iv
Professional Engineers' Examination: *General Information and Suggestions to Candidates*	v

Part I: AM Sample Exams and Solutions

AM SAMPLE EXAM 1	1
AM SAMPLE EXAM 2	15
SOLUTIONS-AM SAMPLE EXAM 1	29
SOLUTIONS-AM SAMPLE EXAM 2	48

Part II: PM Sample Exams and Solutions

PM SAMPLE EXAM 1	68
PM SAMPLE EXAM 2	82
PM SAMPLE EXAM 3	92
PM SAMPLE EXAM 4	104
PM BONUS EXAM	117
SOLUTIONS- PM SAMPLE EXAM 1	123
SOLUTIONS- PM SAMPLE EXAM 2	140
SOLUTIONS- PM SAMPLE EXAM 3	153
SOLUTIONS- PM SAMPLE EXAM 4	171
SOLUTIONS-PM BONUS EXAM	191

Preface

The new format of the 'principles and practice of chemical engineering examination will be effective beginning with April 2000 examination. The examination will consist of a total of 80 multiple-choice questions. 40 questions are to be set in each of the morning and afternoon portions of the exam. The Sample Questions and Solutions published by NCEES indicate the allocation of problems to various subjects as follows:

Subject	Approximate % of exam	calculated number of AM and PM problems
Mass & Energy Balance	20	8
Heat Transfer	15	6
Fluids	15	6
Thermodynamics	10	4
Mass Transfer	15	6
Kinetics	10	4
Plant design	15	6
	100	40

Of the seven topics listed above, the first six cover the standard chemical engineering concepts of those subjects such as material and energy balances, industrial heat transfer including heat exchanger design, conduction, evaporation, Bernoulli's equation, piping network problems, control valves, pumps and compressors, laws of thermodynamics, estimation of physical properties, combustion, equilibria and refrigeration. Kinetics covers interpretation and modeling of reaction rate data, reactor design, reaction control, and comparison of reactor types. The subject of mass transfer will encompass typical unit operations of chemical engineering such as absorption, distillation, liquid-liquid extraction, leaching; humidification and dehumidification, drying and crystallization.

The subject of plant design is very comprehensive and will cover various topics such as design optimization, safety, environmental and waste treating, solids separation, vapor-liquid separations, flow sheets, HAZOP analysis, fault tree analysis, scheduling techniques, material selection, equipment sizing and fabrication, life cycle cost, process control including sensors, transmitters and controllers, control loops, and simulation. It will also cover material science as concerned with physical, chemical and thermodynamic properties of matter, strength of materials, phase and equilibrium diagrams, PVT data and relationships, heats of vaporization and molecular structure. Some problems will require application of engineering economics.

As is clear from the above listing, the scope of the PE examination is very broad. Judging from the sample questions, the exams will consist of sets of questions. Each set will have a main problem statement followed by multiple choice one to six questions. In the present booklet of practice examinations, two morning and four afternoon exam sets are presented. The questions are set to cover most of the subject topics outlined by NCEES so as to give the reader a quick and extensive review of the subject material for the PE examination. The format in which these exams are presented may not correspond to the one of NCEES Sample Questions & Solutions exactly. However, it is broad enough for the preparation of the exam. For those taking the new Breadth/Depth exam, the two AM exams cover all of the areas of the Breadth exam, and the four PM exams cover those topics of the Group I, II, and II options of the afternoon exam.

The PE examination is open-book. The candidate may use standard textbooks, handbooks, bound reference material and a battery-operated calculator. However, many boards do not allow use of aid-books such as this one and even review books containing many solved problems. Best way to prepare for the exam is to have a thorough review of the principles and extensive practice of solving problems and not depend upon referencing material from text books or handbooks as this takes up too much of valuable time which can otherwise be used to solve the problems. Referencing should be done only when the solution of the problem implicitly requires it

I wish good luck to my readers in their P.E. examinations and every success in their professional careers.

Rajaram K. Prabhudesai

General Information and Suggestions to Candidates

Becoming A Professional Engineer

To achieve registration as a Professional Engineer there are four distinct steps: education, fundamentals of engineering (engineer-in-training) exam, professional experience, and finally, the professional engineer exam. These steps are described in the following sections.

Education
The obvious appropriate education is a B.S. degree in chemical engineering from an accredited college or university. This is not an absolute requirement. Alternative, but less acceptable, education is a B.S. degree in something other than chemical engineering, or from a non-accredited institution, or four years of education but no degree.

Fundamentals of Engineering (FE/EIT) Exam
Most people are required to take and pass this eight-hour multiple-choice examination. Different states call it by different names (Fundamentals of Engineering, E.I.T., or Intern Engineer) but the exam is the same in all states. It is prepared and graded by the National Council of Examiners for Engineering and Surveying (NCEES). Review materials for this exam are found in other books like Newnan: Engineer-In-Training License Review and Newnan and Larock: Engineer-In-Training Examination Review.

Experience
Typically one must have four years of acceptable experience before being permitted to take the Professional Engineer exam, but this requirement may vary from state to state. Both the length and character of the experience will be examined. It may, of course, take more than four years to acquire four years of acceptable experience.

Professional Engineer Exam
The second national exam is called Principles and Practice of Engineering by NCEES, but probably everyone else calls it the Professional Engineer or P.E. exam. All states, plus Guam, the District of Columbia, and Puerto Rico use the same NCEES exam.

Chemical Engineering Professional Engineer Exam Background

The reason for passing laws regulating the practice of chemical engineering is to protect the public from incompetent practitioners. Beginning about 1907 the individual states began passing title acts regulating who could call themselves a chemical engineer. As the laws were strengthened, the practice of certain aspects of chemical engineering was limited to those who were registered chemical engineers, or working under the supervision of a registered chemical engineer. There is no national registration law; registration is based on individual state laws and is administered by boards of registration in each of the states.

Examination Development
Initially the states wrote their own examinations, but beginning in 1966 the NCEES took over the task for some of the states. Now the NCEES exams are used by all states. This greatly eases the

ability of a chemical engineer to move from one state to another and achieve registration in the new state. About 1600 chemical engineers take the exam each year. As a result about 23% of all chemical engineers are registered professional engineers.

The development of the chemical engineering exam is the responsibility of the NCEES Committee on Examinations for Professional Engineers. The committee is composed of people from industry, consulting, and education, plus consultants and subject matter experts. The starting point for the exam is a chemical engineering task analysis survey that NCEES does at roughly five to ten year intervals. People in industry, consulting and education are surveyed to determine what chemical engineers do and what knowledge is needed. From this NCEES develops what they call a "matrix of knowledge" that form the basis for the chemical engineering exam structure described in the next section.

The actual exam questions are prepared by the NCEES committee members, subject matter experts, and other volunteers. All people participating must hold professional registration. Using workshop meetings and correspondence by mail, the questions are written and circulated for review. The problems relate to current professional situations. They are structured to quickly orient one to the requirements, so the examinee can judge whether he or she can successfully solve it. While based on an understanding of engineering fundamentals, the problems require the application of practical professional judgement and insight. While four hours are allowed for forty problems, probably any problem can be solved in 5 minutes by a specialist in the field. A professionally competent applicant can solve the problem in no more than 6 minutes. Multi-part questions are arranged so the solution of each succeeding part does not depend on the correct solution of a prior part. Each part will have a single answer that is reasonable.

Examination Structure

The forty problems in the morning four-hour session are multiple-choice. In the afternoon four-hour session all forty problems are multiple choice. In each category (Thermodynamics, Process design, Mass transfer, and so on) about half of the problems will be in the morning session and half in the afternoon. Engineering economics may appear as a component within one or two of the problems.

Note: The examination is developed with problems that will require a variety of approaches and methodologies including design, analysis, application, economic aspects, and operations.

Taking The Exam

Exam Dates

The National Council of Examiners for Engineering and Surveying (NCEES) prepares Chemical Engineering Professional Engineer exams for use on a Friday in April and October each year. Some state boards administer the exam twice a year in their state, while others offer the exam once a year. The scheduled exam dates are:

	April	October
2000	14	27
2001	20	26
2002	19	25
2003	11	24
2004	16	29

People seeking to take a particular exam must apply to their state board several months in advance.

Exam Procedure

Before the morning four-hour session begins, the proctors will pass out an exam booklet and solutions pamphlet to each examinee. There are likely to be civil, electrical, and mechanical engineers taking their own exams at the same time. You must solve four of the ten chemical engineering problems.

The solution pamphlet contains grid sheets on right-hand pages. Only work on these grid sheets will be graded. The left-hand pages are blank and are for scratch paper. The scratch work will not be considered in the scoring.

If you finish more than 30 minutes early, you may turn in the booklets and leave. In the last 30 minutes, however, you must remain to the end to insure a quiet environment for all those still working, and to insure an orderly collection of materials.

The afternoon session will begin following a one-hour lunch break. The afternoon exam booklet will be distributed along with an answer sheet. The booklet will have forty multiple choice questions. You must select and solve all of them. In the Breadth/Depth exam, there will be multiple afternoon exams, you must choose one of the afternoon exams. An HB or #2 pencil is to be used to record your answers on the scoring sheet.

Exam-Taking Suggestions

People familiar with the psychology of exam-taking have several suggestions for people as they prepare to take an exam.

1. Exam taking is really two skills. One is the skill of illustrating knowledge that you know. The other is the skill of exam-taking. The first may be enhanced by a systematic review of the technical material. Exam-taking skills, on the other hand, may be improved by practice with similar problems presented in the exam format.

2. Since there is no deduction for guessing on the multiple choice problems, an answer should be given for all ten parts of the four selected problems. Even when one is going to guess, a logical approach is to attempt to first eliminate one or two of the five alternatives. If this can be done, the chance of selecting a correct answer obviously improves from 1 in 4 to, say, 1 in 3.

3. Plan ahead with a strategy. Which is your strongest area? Can you expect to see one or two problems in this area? What about your second strongest area? What will you do if you still must find problems in other areas?

4. Have a time plan. How much time are you going to allow yourself to initially go through the entire forty problems and grade them in difficulty for you to solve them? Consider assigning a letter, like A, B, C and D, to each problem. If you allow 15 minutes for grading the problems, you might divide the remaining time into five parts of 45 minutes each. Thus 45 minutes would be scheduled for the first - and easiest - problem to be solved. Three additional 45 minute periods could be planned for the remaining problems. Finally, the last 45 minutes would be in reserve. A time plan is very important. It gives you the confidence of being in control, and at the same time keeps you from making the serious mistake of misallocation of time in the exam.

EXAMINATION SPECIFICATIONS
NCEES PRINCIPLES AND PRACTICE EXAMINATION
IN THE DISCIPLINE OF
CHEMICAL ENGINEERING

EFFECTIVE BEGINNING WITH APRIL 2000 EXAMINATION

	Number of Problems
A. FLUID MECHANICS	12

piping network problems; pump sizing or pump performance; compressor sizing or compressor performance; control valve selection problems; fluid flow through beds; two-phase flow; Bernoulli equation applications.

B. HEAT TRANSFER	12

industrial heat transfer including but not limited to the following: heat exchanger design and performance; energy conservation; conduction, especially insulation problems; convection; radiation, especially furnace design; evaporation.

C. CHEMICAL KINETICS	8

interpretation of experimental data and reaction rate modeling; commercial reactor design from rate model and/or product distribution; comparison of reactor types; reaction control.

D. MASS AND ENERGY BALANCES	16

process stoichiometry and material balances; process energy balances; conservation laws.

E. MASS TRANSFER	12

typical applications including but not limited to the following: gas absorption and stripping; distillation; liquid-liquid extraction and leaching; humidification and dehumidification; drying.

F. PLANT DESIGN	12

process and equipment design including but not limited to the following: optimization of design; general safety considerations; environmental and waste treating; solids separation; vapor-liquid separations; flow sheets; HAZOP (hazard and operational) analysis; fault tree analysis; scheduling techniques; sizing and fabrication of equipment; material selection; life cycle cost; physical and chemical properties of matter; strength of materials; crystallographic structure; phase diagrams (metallurgical); latent heat; PVT data and relationships; molecular structure; sensors; transmitters and controllers; control loops; simulation.

G. THERMODYNAMICS	8

estimation and correlation of physical properties; chemical equilibrium; heats of reaction; application of first and second laws; vapor-liquid equilibrium; combustion; refrigeration.

Total number of problems = 80

Note: The examination is developed with problems that will require a variety of approaches and methodologies including design, analysis, application, and operations. Some problems may include engineering economics analyses.

5. Read all four multiple choice answers before making a selection. The first answer in a multiple choice question is sometimes a plausible decoy - not the best answer.

6. Do not change an answer unless you are absolutely certain you have made a mistake. Your first reaction is likely to be correct.

7. Do not sit next to a friend, a window, or other potential distractions.

Exam Day Preparations
There is no doubt that the exam will be a stressful and tiring day. This will be no day to have unpleasant surprises. For this reason we suggest that an advance visit be made to the examination site. Try to determine such items as:

1. How much time should I allow for travel to the exam on that day? Plan to arrive about 15 minutes early. That way you will have ample time, but not too much time. Arriving too early, and mingling with others who also are anxious, will increase your anxiety and nervousness.

2. Where will I park?

3. How does the exam site look? Will I have ample work space? Where will I stack my reference materials? Will it be overly bright (sunglasses) or cold (sweater), or noisy (earplugs)? Would a cushion make the chair more comfortable?

4. Where is the drinking fountain, lavatory facilities, pay phone?

5. What about food? Should I take something along for energy in the exam? A bag lunch during the break probably makes sense.

What To Take To The Exam
The NCEES guidelines say you may bring the following reference materials and aids into the examination room for your personal use only:
1. Handbooks and textbooks
2. Bound reference materials, provided the materials are and remain bound during the entire examination. The NCEES defines "bound" as books or materials fastened securely in its cover by fasteners which penetrate all papers. Examples are ring binders, spiral binders and notebooks, plastic snap binders, brads, screw posts, and so on.
3. Battery operated, silent non-printing calculators.

At one time NCEES had a rule that did not permit "review publications directed principally toward sample questions and their solutions" in the exam room. This set the stage for restricting some kinds of publications from the exam. <u>State boards may adopt the NCEES guidelines, or adopt either more or less restrictive rules</u>. Thus an important step in preparing for the exam is to know what will - and will not - be permitted. We suggest that if possible you obtain a written copy of your state's policy for the specific exam you will be taking. Recently there has been considerable confusion at individual examination sites, so a copy of the exact applicable policy will not only allow you to carefully and correctly prepare your materials, but also will insure that the exam proctors will allow all proper materials.

Clothes - Plan to wear comfortable clothes. You probably will do better if you are slightly cool.
Box For Everything - You need to be able to carry all your materials to the exam and have them conveniently organized at your side. Probably a cardboard box is the answer.

Passing The Exam

In the exam you must answer all eighty problems, each worth 1 point, for a total raw score of 80 points. Stated bluntly, you must get 48 of the 80 possible points to pass. The converted scores are reported to the individual state boards in about two months, along with the recommended pass or fail status of each applicant. The state board is the final authority of whether an applicant has passed or failed the exam.

Although there is some variation from exam to exam, the following gives the approximate passing rates:

Applicant's Degree	Percent Passing Exam
Engineering from accredited school	62%
Engineering from non-accredited school	50
Engineering Technology from accredited school	42
Engineering Technology from non-accredited school	33
Non-Graduates	36
All Applicants	48

Although you want to pass the exam on your first attempt, you should recognize that if necessary you can always apply and take it again.

This Book

This book is organized to cover the chemical engineering professional engineer (principles and practice) exam.

NCEES does not allow their problems to be reproduced, so none of the problems in this book came from them. Each one is structured to approximate the scope and difficulty of the actual exam problems you will encounter. The National Council of Examiners for Engineering and Surveying (NCEES), which prepares the chemical engineering examination, calls it an open book examination. Most states accept this and allow applicants to bring textbooks, handbooks and any bound reference materials to the exam. A few states, however, do not permit review publications directed principally toward sample problems and their solutions.

MORNING SAMPLE EXAMINATION

Instructions for morning Session

1. You have four hours to work on the morning session. Do not write in this handbook.

2. Answer all forty questions for a total of forty answers. There is no penalty for guessing.

3. Work rapidly and use your time effectively. If you do not know the correct answer, skip it and return to it later.

4. Some problems are presented in both metric and English units. Solve either problem.

5. Mark your answer sheet carefully. Fill in the answer space completely. No marks on the workbook will be evaluated. Multiple answers receive no credit. If you make a mistake, erase completely.

Work all 40 problems in four hours.

P.E. Chemical Engineering Exam
Morning Session

Ⓐ Ⓑ Ⓒ Fill in the circle that matches your exam booklet.

M1 Ⓐ Ⓑ Ⓒ Ⓓ	M11 Ⓐ Ⓑ Ⓒ Ⓓ	M21 Ⓐ Ⓑ Ⓒ Ⓓ	M31 Ⓐ Ⓑ Ⓒ Ⓓ
M2 Ⓐ Ⓑ Ⓒ Ⓓ	M12 Ⓐ Ⓑ Ⓒ Ⓓ	M22 Ⓐ Ⓑ Ⓒ Ⓓ	M32 Ⓐ Ⓑ Ⓒ Ⓓ
M3 Ⓐ Ⓑ Ⓒ Ⓓ	M13 Ⓐ Ⓑ Ⓒ Ⓓ	M23 Ⓐ Ⓑ Ⓒ Ⓓ	M33 Ⓐ Ⓑ Ⓒ Ⓓ
M4 Ⓐ Ⓑ Ⓒ Ⓓ	M14 Ⓐ Ⓑ Ⓒ Ⓓ	M24 Ⓐ Ⓑ Ⓒ Ⓓ	M34 Ⓐ Ⓑ Ⓒ Ⓓ
M5 Ⓐ Ⓑ Ⓒ Ⓓ	M15 Ⓐ Ⓑ Ⓒ Ⓓ	M25 Ⓐ Ⓑ Ⓒ Ⓓ	M35 Ⓐ Ⓑ Ⓒ Ⓓ
M6 Ⓐ Ⓑ Ⓒ Ⓓ	M16 Ⓐ Ⓑ Ⓒ Ⓓ	M26 Ⓐ Ⓑ Ⓒ Ⓓ	M36 Ⓐ Ⓑ Ⓒ Ⓓ
M7 Ⓐ Ⓑ Ⓒ Ⓓ	M17 Ⓐ Ⓑ Ⓒ Ⓓ	M27 Ⓐ Ⓑ Ⓒ Ⓓ	M37 Ⓐ Ⓑ Ⓒ Ⓓ
M8 Ⓐ Ⓑ Ⓒ Ⓓ	M18 Ⓐ Ⓑ Ⓒ Ⓓ	M28 Ⓐ Ⓑ Ⓒ Ⓓ	M38 Ⓐ Ⓑ Ⓒ Ⓓ
M9 Ⓐ Ⓑ Ⓒ Ⓓ	M19 Ⓐ Ⓑ Ⓒ Ⓓ	M29 Ⓐ Ⓑ Ⓒ Ⓓ	M39 Ⓐ Ⓑ Ⓒ Ⓓ
M10 Ⓐ Ⓑ Ⓒ Ⓓ	M20 Ⓐ Ⓑ Ⓒ Ⓓ	M30 Ⓐ Ⓑ Ⓒ Ⓓ	M40 Ⓐ Ⓑ Ⓒ Ⓓ

FLUIDS

Bernoulli's equation:

M1: A solution of specific gravity 1.84 is being discharged from a tank to the atmosphere. Level of the liquid in the tank is 20 ft above the centerline of the exit pipe. Frictional and contraction losses in the pipe amount to 12 ft head of solution.

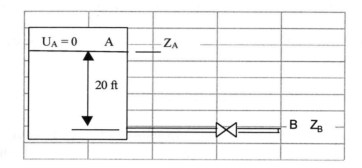

Under these conditions, the discharge velocity [ft/s] from the pipe is closest to:

    (A) 35.2   (B) 7.4   (C) 22.7   (D) 27.8

Control of flow systems:

M2: A manufacturer lists the following C_v data for his line of control valves.

Control valve size, in	1	1½	2	3	4	6	8
Valve coefficient, C_v	45	125	165	350	775	1000	2000

The service conditions of a certain process are as follows:

Fluid	Aqueous solution
Flow rate GPM min / max	125 / 1000
Flow rate Normal GPM @ F. T.	400
Pressure psig max. in / normal out	50 / 25
ΔP psi max / design	30 / 4
Temperature ºF max / normal	100 / 75
Sp. gravity @ 60 ºF / at F.T.	N/A / 1.2

The size [in] of a suitable control valve for this service will be

    (A) 3   (B) 6   (C) 4   (D) 8

Corrosion:

M3: Saponification of fatty acids is carried out in kettles using 12.6 to 14 % NaOH solution. Heat is supplied by direct steam. Kettle is kept boiling for 4 hours. After 4 hours, NaCl is added, and boiling continued another 4 hours. Soap separates as upper layer, and glycerin and NaCl separate as lower layer. The most economic construction material to carry out this operation is

(A) Carbon steel
(B) Stainless steel 304
(C) Glass lined steel
(D) Stainless steel 316

Economics:

M4: A newly installed piece of equipment has a value of $50,000. Its useful life is estimated to be 10 years and its salvage value is $8,000. Depreciation will be charged as a cost by making equal yearly payments, first payment being made at the end of the first year. The depreciation fund will be accumulated at an annual interest rate of 8¼ %. The yearly depreciation cost [$] under these conditions is close to

(A) 4200   (B) 2865   (C) 2300   (D) 3200

Flow in pipes and fittings:

M5: Oil is flowing in a 2" nominal diameter pipe 400 ft long. The data on the pipe are as follows:

Pipe ID = 2.067" = 0.17225 ft.
Inside area of cross section = 0.0233 ft²
Relative roughness of pipe = 0.0009
Density of oil = 51 lb/ft³, Viscosity of oil = 2.2 cP.

If the pressure drop in this line is not to exceed 7.5 psi, the allowable velocity [ft/s] through the pipe is close to:

(A) 6.0   (B) 3.0   (C) 4.7   (D) 8.0

Sensors:

M6: Water is flowing in a 4" schedule 40 pipe. The flow is measured by an orifice of diameter 5 cms. The differential is measured with an electronic differential transmitter which is calibrated 0 to 250 cms of water column and 4 to 20 mA dc output. If the signal from the transmitter is 18.5 mA dc, the flow rate [m³/h] is most nearly:

(A) 48.6   (B) 29.7   (C) 36.2   (D) 42.4

Packed and fluidized beds:

M7: For small particles, the minimum fluidization velocity can be calculated by the equation:

$$u_{mf} = \frac{(\phi_s \bar{d}_p)^2}{150} \left(\frac{(\rho_s - \rho_g)g}{\mu} \right) \left(\frac{\epsilon_{mf}^3}{1 - \epsilon_{mf}} \right)$$

where u_{mf} = Minimum fluidization velocity, cm/s

ϕ_s = Sphericity of particles
$\bar{d}_p$ = particle diameter, cm
ρ_s = particle density, g/cm³
ρ_g = gas density, g/cm³
μ = fluid viscosity, g/(cm.s)
g = acceleration due to gravity, cm/s²
ϵ_{mf} = bed voidage at minimum fluidization velocity

A bed comprising of particles having a mean diameter of 100 μ is to be fluidized with air. The following additional data about the bed are available:

μ = 0.018 cps ρ_s = 1.2 g/cm³ ϕ_s = 0.67 ϵ_{mf} = 0.58 ρ_g = 1.2 x 10⁻³

The minimum fluidization velocity [cm/s] required to fluidize this bed is nearly

(A) 0.44   (B) 1.0   (C) 0.7   (D) 1.3

HEAT TRANSFER

Resistance:

M8: The composite walls of a furnace are made of 9" Kaolin fire brick, 4" insulating brick, and 8" building brick. The mean thermal conductivities of the materials are as follows:

Material	Thermal conductivity Btu/(h.ft².°F/ft)
Kaolin fire brick	0.68
Insulating brick	0.15
Building brick	0.40

When the inside surface of the furnace wall is maintained at 2000°F, the outside surface temperature of the wall was found to be 180 °F. The interface temperature [°F] between the kaolin and insulating bricks will most nearly be

(A) 1592   (B) 1500   C) 1700   (D) 1800

Conduction:

M9 to M10: The composite wall of a furnace is constructed of 6" thick fire brick and common brick of certain thickness. Thermal conductivities of the fire brick and common brick are 0.08 and 0.8 Btu/(h.ft².°F/ft) respectively. The temperature of the inner surface of the wall is 1500 °F and the outer surface of the common brick is at 180 °F. The heat loss from the furnace is found to be 186.4 Btu/h.ft².

M9: The thickness [inches] of the common brick wall is nearly

(A) 6   (B) 9   (C) 8   (D) 4.5

M10: The temperature [°F] at the interface of the fire brick and the common brick is nearly

(A) 840   (B) 335   (C) 515   (D) 680

Energy conservation:

M11: A boiler is producing 5000 lb/h of steam at 150 psig and 400 °F. Water is fed to the boiler at a temperature of 80 °F. The boiler is operating at 80 % efficiency. If the feed water is heated to 220 °F, by using a hot waste stream, % energy savings would be near to

(A) 10.2   (B) 15.1   (C) 9.6   (D) 12.8

Evaporation:

M12 to M13: A single effect evaporator is to concentrate 30,000 lb/h of caustic solution from 10 to 50 % concentration. The evaporator is to be supplied with 30 psia steam. The feed is at 100 °F. It operates at a vacuum 26" Hg. Some additional data are given below

Overall coefficient of heat transfer = 450 Btu/h.ft².°F
Boiling point of caustic solution = 198 °F
Specific heat of steam, C_P = 0.46 Btu/lb.°F

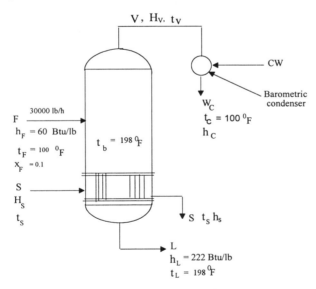

M12: Steam economy [lb evaporated per lb of steam] is nearer to

(A) 1.0   (B) 0.5   (C) 0.837   (D) 0.75

M13: Cooling water required (gpm) is nearer to

(A) 1200   (B) 1100   (C) 1500   (D) 1300

Heat transfer in packed beds:

M14: Bunnell correlated his data on cooling of air passing through a packed bed and obtained the following dimesionless equation for effective thermal conductivities of fluid-solid systems involving spherical particles

$$\frac{k_e}{k_g} = \frac{1}{D_t}\left(\frac{k_p}{k_g}\right)^{0.12}\left[3.4 + 0.00584\frac{D_p G}{\epsilon \mu}\right]$$

where k_e, k_g etc. are variables as described in data given below.

A tube (ID = 3.026") is filled with spherical glass particles to a depth of 12". The following data about the properties of the particles are available

D_p = Diameter of particle = 0.0015 ft
ρ_p = Absolute density of particles = lb/ft³
ϵ = Void fraction = 0.42
μ = Viscosity of fluid (air) flowing through the bed = 0.0435 lb/(h.ft)
k_p = Thermal conductivity of particles = 0.63 Btu/(h.ft².°F)/ft
k_g = Thermal conductivity of medium (air) = 0.0152 Btu/(h.ft².°F)/ft
k_e = Effective thermal conductivity, Btu/(h.ft².°F)/ft
D_t = Diameter of tube or vessel which is packed with solids = 3.026"= 0.2522 ft

The effective thermal conductivity [Btu/(h.ft².°F)/ft] of this solid-air system is close to

(A) 0.4021   (B)   0.2432   (C) 0.3067   (D) 0.3656

Fouling:

M15 to M16: A double pipe countercurrent heat exachager (Outside surface area = 200 ft²) has reached its design capacity of 490,000 Btu/h and is to be shut down for cleaning. The data for the exchanger are as follows:

		Shell side	tube side
Temperature of hot fluid, °F in		350	
Temperature of hot fluid, °F out		450	
Temperature of cold fluid, °F in			300
Temperature of cold fluid, °F out			310
Heat transfer coefficient, Btu/h.ft².°F		38.4	300

Tubes are 1" OD 18 BWG.

M15: The dirty overall coefficient of heat transfer [Btu/h.ft².°F] is nearer to

(A) 50   (B) 40   (C) 28   (D) 35

M16: If the metal wall resistance is negligible and the tube inside dirt factor is 0.003 h.ft².°F/Btu, the dirt factor [h.ft².°F/Btu] on the shell side is close to

(A) 0.003   (B) 0.00265   (C)   0.0025   (D) 0.00285

KINETICS

Control of reactors:

M17 to M18: A first order irreversible reaction $A \rightarrow B$ is carried out in a stirred tank reactor. By a consideration of unsteady material balance over the reactor, the following relation is obtained:

$$V\frac{dC_A}{dt} + (F + kV)C_A = FC_{Ao}$$

where   V = Hold up volume of reaction mass, ft³
        k = First order reaction rate constant, h⁻¹
        C_A = Concentration of species A in the reactor or exit concentration, mols/ft³
        C_{ao} = Feed concentration, mols/ft³
        F = Feed rate, ft³/h

In the derivation of the above equation, it is assumed that both V and ρ, the density of solution are constant.

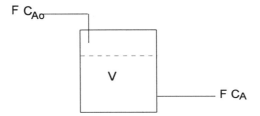

M17: If the hold up time is 1.6 h and the reaction rate constant is 2 h⁻¹, the time constant T of the reactor is near to

(A)   0.28   (B)   0.381   (C)   0.34   (D)   0.52

M18: The transfer function for the CSTR in terms of deviation variables can be derived as follows

$$\frac{C_1(s)}{C_{Ao}(s)} = \frac{K}{Ts + 1}$$

where K = Process gain = $\frac{F}{F + kV}$
    T = Time constant, h
    C_1 = Deviation variable = $C_A - C_{As}$
    C_{ao} = Deviation variable = $C_{ao} - C_{aos}$
    C_{as}, C_{Aos} = Steady state concentrations, mol/ft³

If the concentration of the feed is suddenly changed from 0.75 to 0.72 mols/ft³, the outlet concentration (mols/ft³) after 2 hours from the start of the change would be near to

(A)   0.28   (B)   0.38   (C)   0.17   (D)   0.52

MASS AND ENERGY BALANCE

Conservation of energy:

M19: An endothermic reaction is carried out in a jacketed reaction vessel with initial charge of 500 lbs consisting only of reactants. The feed charge temperature is 68 °F. The reaction is carried out for 2 hours. Heat is supplied to the reactor by condensing saturated steam at 450 °F in the jacket. The reaction mass absorbs heat at 991 Btu/lb of charge during the reaction and heat losses are 8000 Btu/h of reaction time. At the end of 2 hours, the temperature of reaction mass is 212 °F. Under these conditions, the steam usage [lb/batch] is near to

  (A) 772   (B) 824   (C) 916   (D) 863

Conservation of mass:

M20 to M21: A natural gas analyzes 90 % CH_4 and 10 % N_2. It is burned under a boiler and the CO_2 is scrubbed out from the flue gases for the production of dry ice. The analysis of the gas leaving the scrubber shows CO_2 1.1 %, O_2 5 %, and N_2 93.9 %.

M20: The percent CO_2 absorbed in the scrubber is

  (A) 88.8   (B) 95.2   (C) 76.4   (D) 85.7

M21: The percent excess air used is

  (A) 20   (B) 22.5   (C) 25.4   (D) 28.6

Economics:

M22: The installed cost of a machine is $20000. Its salvage value after 10 years of useful life is $4000. Interest rate is 10 % and the annual maintenance cost is $ 1000. The annual cost [$] of the machine is most nearly

  (A) 3004   (B) 3618   (C) 4004   (D) 4506

Mass balance with and without reactions:

M23 to M24: In Haber process to manufacture NH_3, a mixture of nitrogen and hydrogen which contains some argon as impurity is passed through a catalyst at 800 to 1000 atm., and at a temperature of from 500 to 600 °C. NH_3 produced in the reactor is separated from the reaction gases in a separator. NH_3 free gases after a small purge to prevent accumulation of argon are returned to the reactor along with the feed. Product NH_3 does not contain any dissolved gases. The compositions of some streams are indicated in the following figure.

$$N_2 + 3H_2 \rightarrow 2NH_3$$

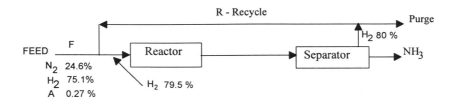

M23: Per 100 moles of fresh feed, the moles recycled are

(A) 932   (B) 880   (C) 738   (D) 804

M24: The percent conversion of hydrogen per pass is near to

(A) 12.23   (B) 9.2   (C) 11.4   (D) 10.8

MASS TRANSFER

Forces and fluxes:

M25: Oxygen (component A) is diffusing through stagnant carbon monoxide (B) under steady state conditions. Carbon monoxide is non-diffusing. The total pressure is 1 atm. and the temperature is 32 °F. The partial pressures of oxygen at two parallel planes 0.084" apart are 0.13 and 0.065 atm. respectively. The diffusivity of oxygen in the mixture is 0.7244 ft²/h. The rate of diffusion of oxygen [lb mol/h.ft²] through the gas mixture is near to

(A) 0.0303   (B) 0.042   (C) 0.028   (D) 0.036

Liquid- solids systems:

M26: The following equation describes the constant pressure drop operation of a filter.

$$(V+V_1)^2 = \frac{2A^2 \Delta P(1-mx)}{\mu \rho x a}(\theta + \theta_1)$$

where V = Volume of filtrate, collected
   V_1 = Volume of filtrate corresponding to a fictitious cake formed due to resistance of filter medium.
   A = Area of filter
   ΔP = Pressure drop
   m = Ratio of wet cake/dry cake
   x = Mass fraction of solids in slurry
   μ = Viscosity of filtrate
   a = Specific resistance of cake
   θ = time of filtration
   θ_1 = *Fictititious time* corresponding to the resistance of the filter medium.

Filtration tests with a certain slurry showed that its specific resistance a is 160 h²/lb. During the tests, the fluid viscosity was 1.1 cP and 3.2 lb of wet cake with 5 % moisture content was formed per cu. ft of filtrate. The filtrate density = 62.4 lb/ft³. The cake is noncompressible and the resistance of filter medium is negligible. If a new unit is to be designed to operate at a constant pressure drop of 10 psi to produce 60 ft³ of filtrate in 40 min, the filter area [ft²] required is near to

(A) 43   (B) 64   (C) 56   (D) 68

PVT data, equilibrium data:

M27: 8 lb of O_2 is vaporized into an initially evacuated vessel of 1.04 ft³ internal volume at -13 °F. The critical constants of oxygen are P_C = 49.7 atm., T_C = -180.4 °F. The pressure [atm.] of oxygen in the cylinder will be near to

(A) 82   (B) 50   (C) 71   (D) 63

Separation systems:

M28 to M29: An equilibrium diagram showing vapor-liquid compositions of heptane at one atm. pressure for the system heptane-ethyl benzene is given below.

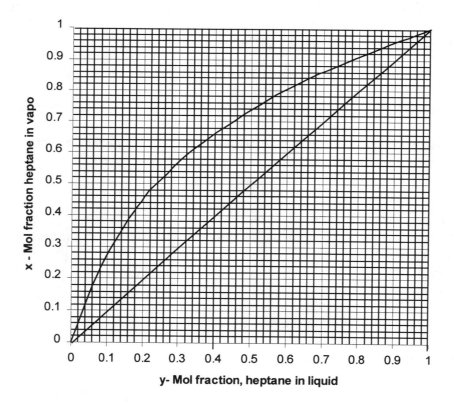

A feed mixture containing 40 mol% heptane and 60 mol% ethyl benzene is to be fractionated at one atm. pressure to produce a distillate containing 97 mol% heptane, and a residue containing 98% ethyl benzene.

M28: If the feed to the column is saturated liquid and the reflux at its bubble point, the minimum reflux ratio required is near to

(A) 0.53 (B) 1.156 (C) 1.34 (D) 1.73

M29: If an actual reflux ratio of 2.5 is used, the theoretical number of equilibrium stages is near to

(A) 10 (B) 8 (C) 9 (D) 11

Transport properties:

M30: Thomas equation to calculate the viscosity of a pure liquid knowing its structure is as follows

$$\log\left(8.569\frac{\mu_L}{\rho^{0.5}}\right) = \theta\left(\frac{1}{T_R} - 1\right)$$

where μ_L = Viscosity of liquid, cP
 ρ_L = Density of liquid, g/cm³
 T_R = Reduced temperature
 θ = A parameter to be calculated from the atomic contributions given in the following table.

Structural contributions to θ

Atom	contribution	Atom	Contribution	
C	-0.462	double bond	0.478	
H	0.249	C$_6$H$_5$	0.385	
O	0.054	S	0.043	
Cl	0.340	CO	0.105	(CO of ketones or esters)
Br	0.326	CN	0.381	(CN of nitriles)

Critical temperature of ethyl acetate (CH$_3$COOC$_2$H$_5$) is 250.1 °C. Its density at 20 °C is 0.901g/cm³. The viscosity of ethyl acetate at 20 °C $\left[\frac{lb}{h.ft}\right]$ is most near to

(A) 0.415 (B) 1.004 (C) 0.415x10^{-2} (D) 0.172

Chemical properties:

M31: The heat of combustion of isopentane [2-methyl butane (C$_5$H$_{12}$)] at 25 °C and 1 atm. is -843.216 kcal/gmol based on water as liquid. The heats of formation of the combustion products are as follows

CO$_2$ (g) $\Delta H_f = -94.05$ *kcal/gmol* at 25 °C and 1 atm.
H$_2$O (l) $\Delta H_f = -68.316$ kcal/gmol at 25 °C and 1 atm.

The heat of formation [kcal/gmol] of isopentane at 25 °C and 1 atm. is near to

(A) - 88.02   (B) -36.93   (C) -84.32   (D) - 63.1

Flooding and pressure drop:

M32 to M33: A packed tower is to be designed to treat 30000 cu ft of gas per hour to remove NH_3 from it. The ammonia content of the gas is 5% by volume. Ammonia-free water will be used as absorbent. Temperature is 68 °F the pressure 15 psia. Ratio of liquid to gas flowrates is 1. 1½" Ceramic Intalox Sadles (F_p = 52) will be used as packing.

M32: If the column is to be designed for a pressure drop of 0.5" H_2O per foot of packing, the diameter [in.] of the column will most nearly be

(A) 18   (B) 30   (C) 24   (D) 36

M33: The percent flood that the column will operate at is most near to

(A) 50   (B) 55   (C) 65   (D) 70

Liquid-gas systems:

M34 to M35: An existing 5.5' diameter tower is to be used to absorb NH_3 from an air stream at 16 psia and a temperature of 80 °F containing 10 mol % NH_3. Feed to the tower will be 1000 lb mols/h. 99% of NH_3 is to be recovered. Ammonia-free water flow rate equal to 1.5 times the minimum will be used. 1" Pall rings (F_p = 56) will be used as packing. The equilibrium relationship over the range of concentrations involved can be represented by the equation

$$y = 1.406x$$

Where y = mol fraction of NH_3 in vapor and x = mol fraction in liquid. The temperature in the tower will be maintained constant at 80 °F by means of cooling coils.

M34: The water rate [gpm/ft².h] to be used is most near to

(A) 5.6   (B) 7.4   (C) 4.2   (D) 6.1

M35: If the overall mass transfer coefficient is 16.2 lbmol/(h.ft².mol fraction), the packed height [ft] needed is most near to

(A) 20   (B) 30   (C) 25   (D) 36

Phase diagrams:

M36: A solution containing 35 % $MgSO_4$ is cooled to 62 °F. During cooling, 2 % of water evaporates. The amount of crystals obtained [lb per ton of original solution] is near to

(A) 1000   (B) 868   (C) 986   (D) 1054

Physical properties:

M37: The Van der Waal's constants for a certain gas are
$$a = 5.1 \times 10^3 \ psia\left[\tfrac{ft^3}{lb\,mol}\right]^2, \qquad b = 0.516 \ \tfrac{ft^3}{lb\,mol}$$

0.0044 lb mol of this gas is contained in a cylinder of internal volume 1.82 ft³. A pressure gage on the cylinder shows a pressure of 5.3 psig. The temperature [°F] of the gas in the cylinder is

(A) 102   (B) 312   (C) 156   (D) 220

PLANT DESIGN

Control:

M38: A control system has the transfer function as given below

$$\frac{Y(s)}{X(s)} = \frac{\tau_1 s + 1}{\tau_2 s + 1}$$

A unit step change is applied to the system. If $\tau_1/\tau_2 = 5$, the maximum value of y(t) is

(A) 0   (B) 5   (C) 1   (D) 4

Equipment design:

M39: A sieve tray column is to be designed to effect an aromatics separation. The following data are available pertaining to the operation of the fractionating column.

	Vapor	Liquid
Flow rate, lb/h	79000	121500
Vol. flow rate, CFS/(gpm)	76.13	(327)
Density, lb/ft³	0.288	46.4

Surface tension = 18 dynes/cm
tray spacing = 24", weir height = 2", hole id. = ¼", hole area/active area = 15 %,
Area of down-comer = 10 % of tray area.
Also, assume wastage = 7.5 % of total tray cross section.

With the above as basis, the diameter [ft] of the column for the required flows will be near to

(A) 5   (B) 6   (C) 6.5   (D) 7

THERMODYNAMICS

Thermodynamic laws:

M40: Air is compressed from 1 atm and 0 °F ($\bar{H}_I = 210.27$ *Btu/lb*) to 10 atm. and 40 °F ($\bar{H}_o = 218.87$ *Btu/lb*). The exit velocity of air from the compressor is 200 ft/s. Inlet air velocity is negligible. If the load is 5 lb/min. of air, the power [hp] required by the compressor is

        (A) 1.3   (B) 1.1   (C) 1.5   (D) 2.0

Stop. Check your work.

End of AM Exam 1.

MORNING SAMPLE EXAMINATION

Instructions for morning Session

1. You have four hours to work on the morning session. Do not write in this handbook.

2. Answer all forty questions for a total of forty answers. There is no penalty for guessing.

3. Work rapidly and use your time effectively. If you do not know the correct answer, skip it and return to it later.

4. Some problems are presented in both metric and English units. Solve either problem.

5. Mark your answer sheet carefully. Fill in the answer space completely. No marks on the workbook will be evaluated. Multiple answers receive no credit. If you make a mistake, erase completely.

Work all 40 problems in four hours.

P.E. Chemical Engineering Exam
Morning Session

Ⓐ Ⓑ Ⓒ Fill in the circle that matches your exam booklet.

A1 Ⓐ Ⓑ Ⓒ Ⓓ	A11 Ⓐ Ⓑ Ⓒ Ⓓ	A21 Ⓐ Ⓑ Ⓒ Ⓓ	A31 Ⓐ Ⓑ Ⓒ Ⓓ
A2 Ⓐ Ⓑ Ⓒ Ⓓ	A12 Ⓐ Ⓑ Ⓒ Ⓓ	A22 Ⓐ Ⓑ Ⓒ Ⓓ	A32 Ⓐ Ⓑ Ⓒ Ⓓ
A3 Ⓐ Ⓑ Ⓒ Ⓓ	A13 Ⓐ Ⓑ Ⓒ Ⓓ	A23 Ⓐ Ⓑ Ⓒ Ⓓ	A33 Ⓐ Ⓑ Ⓒ Ⓓ
A4 Ⓐ Ⓑ Ⓒ Ⓓ	A14 Ⓐ Ⓑ Ⓒ Ⓓ	A24 Ⓐ Ⓑ Ⓒ Ⓓ	A34 Ⓐ Ⓑ Ⓒ Ⓓ
A5 Ⓐ Ⓑ Ⓒ Ⓓ	A15 Ⓐ Ⓑ Ⓒ Ⓓ	A25 Ⓐ Ⓑ Ⓒ Ⓓ	A35 Ⓐ Ⓑ Ⓒ Ⓓ
A6 Ⓐ Ⓑ Ⓒ Ⓓ	A16 Ⓐ Ⓑ Ⓒ Ⓓ	A26 Ⓐ Ⓑ Ⓒ Ⓓ	A36 Ⓐ Ⓑ Ⓒ Ⓓ
A7 Ⓐ Ⓑ Ⓒ Ⓓ	A17 Ⓐ Ⓑ Ⓒ Ⓓ	A27 Ⓐ Ⓑ Ⓒ Ⓓ	A37 Ⓐ Ⓑ Ⓒ Ⓓ
A8 Ⓐ Ⓑ Ⓒ Ⓓ	A18 Ⓐ Ⓑ Ⓒ Ⓓ	A28 Ⓐ Ⓑ Ⓒ Ⓓ	A38 Ⓐ Ⓑ Ⓒ Ⓓ
A9 Ⓐ Ⓑ Ⓒ Ⓓ	A19 Ⓐ Ⓑ Ⓒ Ⓓ	A29 Ⓐ Ⓑ Ⓒ Ⓓ	A39 Ⓐ Ⓑ Ⓒ Ⓓ
A10 Ⓐ Ⓑ Ⓒ Ⓓ	A20 Ⓐ Ⓑ Ⓒ Ⓓ	A30 Ⓐ Ⓑ Ⓒ Ⓓ	A40 Ⓐ Ⓑ Ⓒ Ⓓ

FLUIDS

Bernoulli's equation:

A1: For measuring liquid level in an inaccessible underground tank, the scheme as shown in the figure below is employed. The manometer fluid has a density of 2.5 g/ml while that of the liquid in the tank is 1.05 g/ml. Air flow is controlled so that the frictional, contraction, and

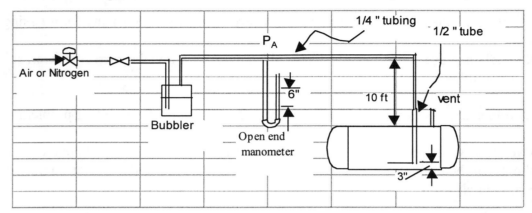

expansion losses in the feed line amount to 1.34" head of liquid in the tank. Velocity in 1/4" tubing is 2.9 ft/s and that in 1/2" pipe is 0.28 ft/s. Density of air = 0.0754 lb/ft^3. If the manometer shows a 6" differential, the level [in] of liquid from the bottom of the tank is close to:

(A) 20   (B) 16   (C) 13   (D) 17

Control of flow systems:

A2: The level of oil in a tank is controlled. Flow range is 100 to 1200 GPM. The oil has a specific gravity of 0.9. Line pressure varies from 100 to 150 psig and the throttling pressure drop varies from 50 to 110 psi. Temperature varies from 70 to 150 °F. The required rangeability of the control valve for this operation is

(A) 8:1   (B) 10:1   (C) 18.0   (D) 25:1

Corrosion:

A3: Wood is digested in the sulfite process in vessels constructed of

(A) Brick lined steel
(B) Stainless steel 304
(C) Glass lined steel
(D) Rubber coated steel.

Economics:
A4: An equipment when installed costs $25,000. Its useful life is 8 years with a salvage value of $5,000. Interest is compounded at 8% per year. The capitalized cost of the equipment is close to

(A) 42,257   (B) 45,000   (C) 40,000   (D) 50,000

Flows in pipes and fittings:

A5: Water is flowing through an annular channel of 50 ft length. The annular channel is made of outer pipe of rectangular tube and an inner schedule 40, ½" pipe. The other data are as follows:

    Outer pipe: 2" square (inside dimensions)
    Inner pipe : OD = 0.84"= 0.07 ft
    Flow area or channel cross section) = 0.02393 ft²
    Relative roughness = 0.0015
    Water flow rate = 100 GPM
    Density of water = 62.4 lb/ft³
    Viscosity of water = 0.9 cP

The frictional pressure drop [psi] for 100 GPM flow in the channel is most nearly:

    (A) 3.2 (B) 6.3 (C) 12.6 (D) 8.4

Sensors:

A6: Air is flowing in a duct 20" id at 220°F and 750 mm pressure. A Pitot tube positioned at center of cross section of the pipe shows a manometer reading of 4" water column. The Pitot tube coefficient is 0.98. Viscosity of air = 0.022 cP. Air flow rate [scfm at 60 °F and 1 atm.] through the duct is nearly

    (A) 16022 (B) 12093 (C) 15000 (D) 10500

HEAT TRANSFER

Packed and fluidized beds:

A7: A certain catalyst consists of microspherical particles having a diameter of 50 μ. The properties of the solid and the air used to fluidize are as follows:

    Solid: $\rho_s = 2.5$ g/cm³, $\phi_s = 1.0$, $\varepsilon_{mf} = 0.56$
    Fluid $\rho_g = 1.2 \times 10^{-3}$ g/cm³, $\mu = 1.8 \times 10^{-4}$ g/(cm.s)

The terminal velocity [cm/s] for these particles is nearly

    (A) 18.9 (B) 25.6 (C) 40.3 (D) 30.7

Resistance:

A8: A hollow metal sphere 4" inside diameter and 2" thick is heated inside so that the inside surface temperature is maintained at 300 °F. If the outside surface temperature of the sphere is maintained at 220 °F, the heat loss [Btu/h] from the sphere is nearly
    (A) 9600 (B) 8720 (C) 6100 (D) 7630

Convection:

A9 to A10: Aniline is maintained at 100 °F in an outdoor storage tank by passing steam through a bundle of tubes immersed in the liquid at the bottom of the tank. The horizontal cylindrical tank is 6' ID by 15' long. The tank is not insulated but shielded from wind. Radiation coefficient can be taken as 0.75 Btu/h.ft².°F. The lowest winter temperature is 0 °F. Steam at 220 °F is passed through the tubes (tube OD = 1"). The properties of aniline as a function of temperature are given in table below

Properties of aniline

t - °F	ρ - lb/ft³	C_p - Btu/lb.°F	μ - cP	k - Btu/h.ft.°F	v ft³/lb
150	61.58	0.4701	1.350	0.09448	0.0162
160	61.28	0.4751	1.221	0.09379	0.0163
170	60.98	0.4797	1.115	0.0931	0.0164

Hints: Kern and other authors have used the following equation to calculate free convection coefficient on the outside of a bank of tubes

$$h_c = 116\left[\left(\frac{k_f^3 \rho_f^2 C_f \beta}{\mu_f'}\right)\left(\frac{\Delta t}{d_o}\right)\right]^{0.25}$$

where h_c = Convection coefficient of heat transfer, Btu/h.ft².°F
k_f = Thermal conductivity at film temperature, Btu/h.ft². °F/ft
ρ_f = density of the fluid, lb/ft³
C_f = specific heat, Btu/lb.°F
β = Coefficient of expansion, 1/°F
μ_f' = viscosity, centipoise
Δt = temperature difference, °F
d_o = Outside diameter of tube, in

Convection coefficient for horizontal plates are given by

For plates facing upward: $h_c = 0.38(\Delta t)^{0.25}$

For plates facing downward: $h_c = 0.20(\Delta t)^{0.25}$

A9: The Grashof number for aniline heating is close to

(A) 6.11x10⁴   (B) 6.11x10⁶   (C) 4.96x10⁶   (D) 3.8x10⁶

A10: Using Kern's equation for a bank of tubes, the convective coefficient by natural convection [Btu/h.ft².°F] is close to

(A) 60   (B) 100   (C) 40   (D) 80

Energy conservation:

A11: The following test data were obtained on a boiler system as a part of energy audit with a view to find avenues for improvement.

Fuel used - natural gas
Flue gas temperature - 500 °F
Steam production = 50000 lb/h at 150 psig and saturated.(No superheat)

	percent
Excess air	90
Stack loss	27.5
Combustion efficiency	72.5
Radiation heat loss	2.0
Blow down heat loss	0.2

The opportunity to improve the efficiency of this operation very substantially lies in

(A) Reduction in flue gas temperature
(B) Reduction in excess air used
(C) Reduction in blow down loss
(D) Reduction in ash losses

Heat transfer in packed beds:

A12: Leva presented the following relation to calculate average heat transfer coefficient h_m for a solid-fluid system

$$\frac{h_m D_t}{k_g} = 0.813 \, e^{-6(D_p/D_t)} \left(\frac{D_p G}{\mu}\right)^{0.9} \text{ for } D_p/D_t < 0.35$$

Where h_m = Mean heat transfer coefficient, Btu/h.ft².°F
D_t = Diameter of tube, ft
k_g = Thermal conductivity of fluid, Btu/h.ft².°F
D_p = diameter of particle, ft
G = Mass velocity of fluid, lb/h.ft²
μ = Viscocity of fluid, lb/h.ft

If the downwards air-flow rate through the bed is 1000 lb/h.ft², the average heat transfer coefficient (Btu/h.ft².°F) is close to

(A) 2.1   (B) 1.144   (C) 3.0   (D) 2.5

Radiation:

A13: Two very large walls are maintained at constant temperatures of 1200 and 1000 °F respectively. Their emissivities are 0.9 and 0.75. The heat [Btu/h.ft²] that need to be removed from the colder wall is close to

(A) 5927   (B) 4236   (C) 3340   (D) 3653

A14: A 2" IPS steel pipe carrying steam at 300 °F is insulated with ½" of rockwool (k = 0.033 Btu/h.ft².°F/ft). The temperature of the wool surface exposed to surrounding air (t = 70 °F) is 125 °F. The emissivity of the insulation is 0.9. The fictitious radiation coefficient of heat transfer [h_r, Btu/h.ft². °F] is nearer to

(A) 2.5   (B) 1.1   (C) 1.5   (D) 1.7

Insulation:

A15: An 1" steel pipe carrying steam at 366 °F will lose heat equal to 306 Btu/h.lin.ft of pipe if it is not insulated. On adding 2" thk. insulation of k = 0.042 Btu/h.ft².°F/ft and emissivity = 0.47, the insulation surface temperature was measured to be 145 °F. The reduction in heat loss [Btu/h.lin.ft of pipe] that is obtained because of addition of the insulation will be nearer to

(A) 155   (B) 120   (C) 96   (D) 135

[You may neglect the resistances of the steam film and the tube wall as they are very small]

A16: A 2" steel pipe is carrying steam at 400 °F. It is covered with an insulation of k = 0.12 Btu/h.ft².°F/ft. The insulation thickness is such that the surface coefficient of heat transfer is 2.3 Btu/h.ft².°F/ft. The critical radius [inches] for this insulation is close to

(A) 0.052   (B) 0.63   (C) 0.9   (D) 0.66

KINETICS

Equilibrium:chemical/phase:

A17 to A18: For the reaction A $\rightleftarrows$ B, the following data are available

at 25 °C, heat of reaction, $\Delta H_r^0 = -23.99$ kcal/gmol

$\Delta G_f^0 = -3.0$ kcal/gmol

For the temperature range of interest, ΔH_r can be assumed to be constant.

A17: The equilibrium constant at a temperature of 50 °C is nearer to

(A) 5.06   (B) 6.9   (C) 158.2   (D) 79.3

A18: If initial concentration of B is zero, the equilibrium conversion [%] at 50 °C is close to

(A) 87.34   (B) 79.1   (C) 67.2   (D) 98.3

MASS AND ENERGY BALANCE

Conservation of energy:

A19: A manufacturer of high quality lime uses precipitated, essentially pure $CaCO_3$ to produce high quality lime. In the process, $CaCO_3$ after heating to 400 °F in a preheater is fed to a rotary kiln. The calcined product leaves the kiln at 1800 °F. Assume 100 % conversion. The following data are available.

Average specific heats

component	Btu/(lb.°F)
$CaCO_3$	0.242
CaO	0.245
CO_2	0.283

Heat of reaction

$$\Delta H_R = 788.4 \text{ Btu/lb of } CaCO_3$$

If the kiln heat efficiency is 80 %, the actual heat [MM Btu/h] to be supplied to the kiln is

(A) 23.4 (B) 29.25 (C) 27.3 (D) 25.2

Conversion/yield:

A20: A plant makes very high grade lime by calcination of pure $CaCO_3$ in a rotary kilns. In one kiln, the reaction goes to 95 % completion. The yield in terms of lb of CO_2 produced per lb of calcium carbonate is

(A) 0.242 (B) 0.418 (C) 0.324 (D) 0.366

A21: Aluminum sulfate is manufactured by treating crushed bauxite ore with sulfuric acid. The ore contains 55.4 % Al_2O_3. Sulfuric acid is 77.7 % H_2SO_4. In one batch, 2160 lb of bauxite ore were treated with 5020 lb of sulfuric acid to obtain a final solution containing 3600 LB of aluminum sulfate. Percent conversion of Al_2O_3 is near to

(A) 84.7 (B) 67.4 (C) 89.6 (D) 96.2

Economics:

A22: An alternate proposal is under consideration to install a device with initial investment of $300,000. The maintenance cost for this machine is $7,000 per year 10 years from now and every 10th year thereafter. In addition, there is $1,000 miscellaneous charge every year. The capitalized cost [$] for this machine is most nearly

(A) 290140 (B) 318540 (C) 346720 (D) 300840

MASS TRANSFER

Cooling towers:

A23 to A24: A cooling tower is operating at steady state under the following conditions.

   Cooling water to the tower = 1500 gpm = 1302 lb/h.ft²
   Temperature of inlet hot water = 120 °F
   Temperature of cooling water out = 85 °F (10 °F approach to wet bulb t = 75 °F)
   Air flow entering at the bottom of tower = 1400 lb/h.ft²
   Inlet air dry bulb temperature = 85 °F, Air wet bulb temperature = 75 °F
   Outlet air: dry bulb temperature = 105 °F, wet bulb temperature = 99 °F
   Tower cross section = 576 ft²

The heat and mass transfer coefficients for the fill used in the tower are given by

   Heat transfer: $h_l a = 0.03(L')^{0.51}(G')$ Btu/h.ft³.°F

   Mass transfer: $k_g a = 0.04(L')^{0.26}(G')^{0.72}$ mols/h.ft³.

A plot of saturation enthalpy vs temperature is made as in the figure below. An operating line

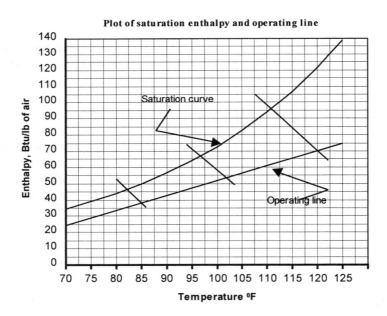

with a slope of $\dfrac{L'C_L}{G'} = \dfrac{1302\,(1)}{1400} = 0.93$ is also drawn.

A23: The number of transfer units, $(NTU)_G$, which the tower is exibiting is near to

(A) 2.5 (B) 3.3 (C) 1.97 (D) 5.4

A24: The humidity of air entering the tower [LB of water vapor/lb dry air] is near to

(A) 0.002 (B) 0.005 (C) 0.0168 (D) 0.1

Forces and fluxes:

A25: Water is flowing down a vertical tube (ID= 3.026", L = 4 ft) at a rate of 125 lb/(h.ft.). The film thickness is 7.94×10^{-4} ft. Re No. is 230. Diffusivity of CO_2 is 7.6×10^{-5} ft²/h. Density of solution can be taken as 62.3 lb/ft³. Solubility of CO_2 in water at 77 °F is 0.002 lbmol/ft³. The mass transfer coefficient $K_{Lav} = \left(\frac{6 D_{AB} \Gamma}{\pi \rho \delta L}\right)^{0.5} = 0.3028 \ \frac{lbmol}{ft^2 . h . \frac{lbmol}{ft^3}}$.

Under these conditions, the rate of diffusion of CO_2 [lbmol/(h.ft of film width] is near to

(A) 3.6×10^{-3} (B) 1.818×10^{-3} (C) 2.4×10^{-3} (D) 3.0×10^{-3}

Gas-solid systems:

A26 to A27: An insulated rotary dryer is to be designed to dry a crystalline material from 30 to 0.5 % moisture. The drier is to be 5 ft in diameter. The moisture in the material can be treated as unbound moisture. Air at 80 °F (dry bulb) and 65 °F (wet bulb) temperature will be heated to 240 °F and fed to the drier countercurrent to the flow of the wet material. The air will leave the dryer at 110 °F dry bulb. The wet solids will enter the dryer at 80 °F and are expected to be discharged at 140 °F. Specific heat of dry solids is 0.3 Btu/(lb.°F). The volumetric heat transfer coefficient for rotary dryers is evaluated by the following equation proposed by Friedman and Marshall

$$Ua = \frac{10 G^{0.16}}{D}$$

Where Ua = Volumetric coefficient of heat transfer, Btu/h.Ft³ °F.

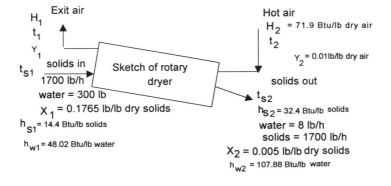

A26: To effect the required drying, the dry air rate [lb/h] to the rotary dryer will be near to

(A) 5000   (B) 7000   (C) 9850   (D) 11000

A27: Assuming no heat loss from the dryer, the heat transfer units [N_{toG}] needed are near to

(A) 1.5   (B) 2.236   (C) 2.0   (D) 1.2

Liquid-solid systems:

A28: 1000 lb of a solid containing 50 % solute and 50 % inerts is to be extracted with solvent C in a two stage crosscurrent extraction using 3000 lb of solvent each time. In each stage, the extracted solids are screw-pressed. The pressed solid contains 1.2 lb of solution per lb of inerts. A triangular diagram is provided below for solving the problem.

Figure for problem A28

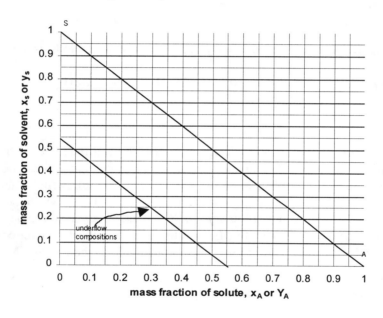

The % recovery of solute in the two stage process is near to

(A) 98   (B) 99   (C) 90   (D) 93

PVT data, equlibrium data:

A29: In case of partially miscible components of a binary system, if the vapor phase behaves as an ideal gas (low pressures), vapor liquid equilibrium relationship can be calculated with the assumption that Raoult's law applies to a component with high concentration. Henry's law is applicable to a component if its concentration is low. If Henry's law holds for one component, then Raoult's law applies to the other component in high concentration.

The following data on two phases is available for water (A) and ether (B) system.

t °C	Total P atm.	saturated water phase x_B	saturated ether phase x_B	P_A atm.	P_B atm.
50	1.744	0.0103	0.9348	0.121	1.679

The mol fraction of ether in vapor in equilibrium with the liquid at 50 °C and 1 atm. total pressure will be near to

(A) 0.12 (B) 0.88 (C) 0.995 (D) 0.93

Transport properties:

A30: An empirical correlation suggested by Fuller et. al. for the calculation of diffusivity of a gas in a binary system is

$$D_{AB} = \frac{10^{-3} T^{1.75} \left[\frac{1}{M_A} + \frac{1}{M_B} \right]^{0.5}}{P \left[(\Sigma v_A)^{\frac{1}{3}} + (\Sigma v_B)^{\frac{1}{3}} \right]^2}$$

Where T is in Kelvin and P in atm. and v_A and v_B are calculated using atomic and molar volumes in the following table.

Atomic and structural diffusion volume

Atomic Diffusion Volumes				Diffusion Volumes for simple molecules			
C	16.5	Cl	19.5	H_2	7.07	CO	18.9
H	1.98	S	17.0	N_2	17.9	CO_2	26.9
O	5.48	Aromatic ring	-20.2	O_2	16.6	NH_3	14.9
N	5.69	Heterocyclic ring	-20.2	Air	20.1	H_2O	12.7
				Cl_2	37.7	SO_2	41.1

The diffusivity [cm²/s] of methyl chloride in air at 80 °C and 1 atm. pressure is most near to

(A) 0.175 (B) 0.175x10⁻⁴ (C) 0.68 (D) 0.13

Chemical properties:

A31: Carbon dioxide reacts with hydrogen according to the following reaction.

$$CO_2(g) + 4H_2(g) = 2H_2O(g) + CH_4(g)$$

The heat of reaction is -39.43 kcal/gmol at 25 °C and 1 atm. Specific heat difference between the products and reactants is given by

$$\Delta C_P = -14.95 + 11.082 \times 10^{-3} T - 1.729 \times 10^{-6} T^2$$

Where T is in K. The heat of reaction [kcal/gmol] at 500 °C is near to

(A) -35.46 (B) - 43.96 (C) - 40.23 (D) - 37.81

Condensation of mixed fluids:

A32 to A33: A mixture of organic vapors at a temperature of 230 °F is to be condensed at 30 psia. The condensing curve was calculated and the results are summarized in the table following.

Temperature, °F	wt. fraction vapor	Enthalpy change MM Btu/h	Heat transferred Btu/h	Δt_W * °F
230	1.0000	- 0.0000		
220	1.0000	- 0.3903	390307	1.22
213.8	1.0000	- 0.6315	241200	0.76
210	0.8425	- 2.4711	1839600	5.75
200	0.4813	- 6.7959	4324800	13.53
190	0.1662	-10.6927	3896800	12.20
185	0.0000	-12.7833	2090600	6.54

* cooling water = 319583 lb/h. Δt_w = *temperature rise of water,* °F

A32: The heat load [MM Btu/h] in the condensation range is near to

(A) 12.152   (B) 12.78   (C) 10.7   (D) 10.06

A33: Weighted temperature difference for the condenser is

(A) 108   (B) 115   (C) 110   (D) 112

Humidification:

A34 to A35: A plant uses acetone as solvent in a process. It is recovered by evaporating it in a stream of nitrogen gas. The mixture of nitrogen and acetone recovered flows through a pipe. At one point in the pipe, the pressure and temperature of the gas are 800 mm Hg and 104 °F. A wet bulb thermometer (wick wetted with acetone) shows a temperature of 80 °F. Additional data to be used for the solution of these problems are given below

Estimated average properties of nitrogen-acetone mixture

Viscosity = 0.0127 cP,
Thermal conductivity mixture = 0.0115 Btu/(h.ft².°F/ft)
Average specific heat = 0.282 Btu/(lb. °F)
Vapor pressure of acetone at 80 °F = 255 mm Hg

Additional properties of acetone

Latent heat of vaporization of acetone at 80 °F = 235 Btu/lb
Diffusivity of acetone in N_2 = 0.523 ft²/s
Specific heat of acetone = 0.318 Btu/lb.°F
Specific heat of nitrogen = 0.248 Btu/lb.°F

A34: Lewis number for this acetone-nitrogen mixture is most near to

(A) 1.2   (B) 0.84   (C) 1.4   (D) 1.12

A35: If nitrogen fed to the recovery unit is acetone-free, the acetone picked up by nitrogen (lb/lb nitrogen) in the recovery unit is most near to

(A) 0.92   (B) 0.75   (C) 0.97   (D) 1.1

Phase diagrams:

A36: The phase diagram for the system Pyridine (C)-water (A)-chlorobenzene (B) is provided below.

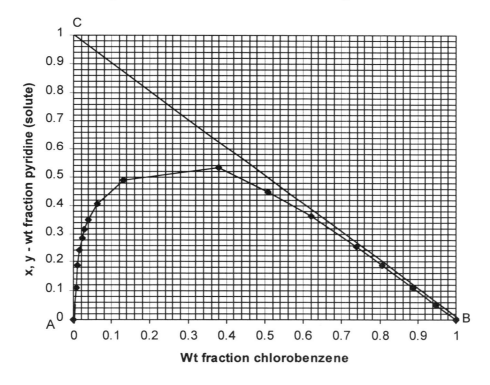

Pyridine-water-chlorobenzene system

2000 lb of a pyridine-water mixture free of chlorobenzene and containing 50 % pyridine is to be extracted in a single batch extraction with chlorobenzene as solvent. The concentration of pyridine in the raffinate is to be reduced from 50 % to 2 % mass fraction. The solvent amount [lb/batch] required is to

(A) 20400   (B) 21867   (C) 23206   (D) 18674

A37: H.W. Prengle et. al. give the following data for n-butane.

t (°F)	P (atm)	V_l (ft³/lb)	V_g (ft³/lb)
260	24.662	0.0393	0.222
270	27.134	0.0408	0.192
280	29.785	0.0429	0.165
290	32.624	0.0458	0.138
305.56	37.47	0.0712	0.0712

The latent heat of vaporization of n-butane [Btu/lb] at 280 °F is near to

    A) 102   (B) 312   (C) 67.5   (D) 220

PLANT DESIGN

Economics:

A38: The total capital investment for a chemical plant is 1.5 million dollars. Working capital is $180,000. The plant can produce 28205 lb of a product per day during a 360 days/yr operating time. The selling price [$/ton] of the product required to give a turn over ratio of 1.0 is near to

    (A) 250   (B) 275   (C) 245   (D) 260

A39: A bond has a maturity value of $3000 and is paying discrete compound interest at an effective annual rate of 5 %. If the bond is sold to a purchaser for $2000, 4 years before its maturity, the effective interest rate [%] the purchaser receives is nearer to

    (A) 20.16   (B) 18.53   (C) 22.26   (D) 16.67

Control:

A40: A control system has transfer functions as shown in the following figure.

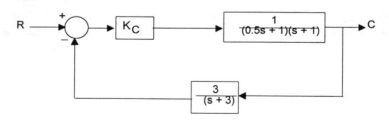

By applying Ruth test, only the following can be said about this control system:

    (A) The system is stable if value of $K_C < 10$
    (B) The system is oscillatory if $6 < K_C < 10$
    (C) The system is stable as well as non-oscillatory if $0 < K_C < 6$
    (D) The system is unstable at all values of K_C

Stop. Check your work. End of AM exam 2.

SOLUTIONS TO AM EXAMS
AM EXAM 1:
FLUIDS

Bernoulli's equation:

M1:

Apply Bernoulli's theorem between points A (surface of the liquid) and B (discharge point). Also take level at B as datum plane.

$$\frac{P_A}{\rho_A} + Z_A + \frac{U_A^2}{2g_c} - F + w = \frac{P_B}{\rho_B} + Z_B + \frac{U_B^2}{2g_c}$$

$$\frac{P_A}{\rho_A} = \frac{P_B}{\rho_B} = \frac{P_{atm}}{\rho} \quad \text{Therefore, these terms cancel out.}$$

$U_A = 0$ since the cross section of the tank is very large.
$w = 0$ since no pump in the line.
$Z_B = 0$ since point B is taken as datum plane.

Therefore, Bernoulli equation reduces to: $\quad Z_A = F + \frac{U_B^2}{2g_c}$

$U_B = \sqrt{2g_c(Z_A - 12)} = \sqrt{2 \times 32.2 \times (20 - 12)} = 22.7$ ft/s (6.92 m/s)

Answer is (C)

Control of flow systems:

M2:

C_{vc} at max. flow $= 1000\sqrt{\frac{1.2}{4}} = 547.7$

C_{vc} at min. flow $= 125\sqrt{\frac{1.2}{30}} = 25.0$

From the list of valves, 4" valve has a maximum flow coefficient of 775.
If this valve is selected,

$$\frac{C_v}{C_{vc}} = \frac{775}{547.7} = 1.415$$

This ratio falls between 1.25 to 2 which is required for good operation of the control valve.

Answer is (C)

Corrosion:

M3:

Saponification is usually carried out in steel vessels since the solution is alkaline and the corrosion rate is not very high. Cheaper material is therefore selected.

Answer is (A)

Economics:

M4:

Total depreciation = S = 50000 - 8000 = $42,000
Yearly payout = R
Number of payments = 10
Annual interest rate I = 8.25 % = 0.0825

$$R = S \frac{i}{(1+i)^n - 1} = 42000 \times \frac{0.0825}{(1+0.0825)^{10} - 1} = 2865 \text{ dollars}$$

Answer is (B)

Flow in pipes and fittings:

M5:

$$\Delta P = \frac{fLu^2\rho}{2g_c d_i (144)}$$

Since both f and u are unknown, solution is by trial and error
For first trial assume u = 4 ft/s

$$\frac{Du\rho}{\mu} = \frac{0.1723 \times 4 \times 51}{2.2 \times 0.000672} = 2.38 \times 10^4 \quad \text{Roughness ratio} = 0.0009$$

f = 0.027 from Moody chart

$$\Delta P = \frac{0.027 \times 400 \times 4^2 \times 51}{2.2 \times 0.000672 \times 144} = 5.52 \text{ psi} \quad \text{This is} < 7.5 \text{ psi required.}$$

Assume u = 4.66 ft/s [7.5/5.52 = $(u_2/u_1)^2$ from which u_2 = 4.66 ft/s]

$$\frac{Du\rho}{\mu} = \frac{4.66}{4.0} \times 2.38 \times 10^4 = 2.8 \times 10^4 \quad f = 0.027 \text{ approx.}$$

$$\Delta P = \frac{0.027 \times 400 \times 4.66^2 \times 51}{64.4 \times (0.1723) \times 144} = 7.49 \text{ } psi$$

which is very close to required value

Therefore u = 4.66 ft/s ≐ 4.7 ft/s

Answer is (C)

Sensors:

M6:

$$\beta = \frac{5}{10.226} = 0.489 \qquad \beta^4 = 0.05716 \qquad 1 - \beta^4 = 0.9428$$

$$h = \frac{18.5 - 4}{20 - 4} \times 250 = 226.6 \ cms$$

Assume flow is fully turbulent, so that orifice coefficient $C_O = 0.61$

$$u_o = 0.61 \sqrt{\frac{2 \times g \times \Delta H}{1 - \beta^4}} = 0.61 \sqrt{\frac{2 \times 9.81 \times 226.6}{0.9428}} = 4.2 \ m/s$$

Check Reynold's number.

$$\frac{D u \rho}{\mu} = \frac{0.05 \times 4.2 \times 1000}{1 \times 10^{-3}} = 3.1 \times 10^6 \quad \text{Flow is turbulent. } C_o = 0.61$$

Flow = $0.7854(0.05)^2 (4.2)(3600) = $ **29.7** m³/h

Answer is (B)

Packed and fluidized beds:

M7:

Solution involves substitution of given values in the equation and completing the arithmetic.

$$\frac{(\phi_s \bar{d}_p)^2}{150} = \frac{(0.67 \times 0.01)^2}{150} = 0.145 \times 10^{-6}$$

$$\frac{(\rho_s - \rho_g)g}{\mu} = \frac{(1.2 - 1.2 \times 10^{-4}) \times 981}{0.00018} = 6660$$

$$g\left(\frac{\epsilon_{mf}^3}{1 - \epsilon_{mf}}\right) = 981\left(\frac{0.58^3}{1 - 0.58}\right) = 455.73$$

$$u_{mf} = 0.145 \times 10^{-6}(6660)(455.73) = 0.44 \ cm/s$$

Answer is (A)

HEAT TRANSFER
Resistance:
M8:
There are 3 resistances in series.

$$R_1 = \frac{9/12}{0.68(1)} = 1.103 \qquad R_2 = \frac{4/12}{0.15(1)} = 2.222 \qquad R_3 = \frac{8/12}{0.4(1)} = 1.667$$

Total resistance = 1.103 + 2.222 + 1.667 = 4.992

Temperature drop is proportional to resistance.
Resistance of Kaolin brick = 1.103
Total temperature drop = 2000 - 180 = 1820 °F
Therefore, the temperature drop across Kaolin brick = $\frac{1.103}{4.922} \times 1820 \simeq 408$ °F
Then interface temperature of Kaolin and insulating brick = 2000 - 408 = **1592** °F

Answer is (A)

Conduction:

M9:

$$q = \frac{\Delta t}{R_T}$$

$R_T = \frac{\Delta t}{q} = \frac{1500 - 180}{186.4} = \frac{1320}{186.4} = 7.082$ h.ft². °F/Btu   total resistance

Resistance of fire brick = $\frac{6/12}{0.08(1\ ft^2)} = 6.25$ h.ft². °F/Btu

Therefore resistance of common brick = 7.082 - 6.25 = 0.832 h.ft². °F/Btu

ΔX of common brick = (0.8)(1)(0.832) = 0.6656 ft = 7.9872" ≅ 8"

Answer is (C)

M10:

Temperature drop across fire brick = (6.25/7.082)(1320) = 1165 °F

Interface temperature between fire brick and common brick = 1500 - 1165 = 335 °F

Answer is (B)

Energy conservation:

M11:

Enthalpy of boiler steam = 1214.7 Btu/lb from steam tables.

Savings in energy = $\frac{(220 - 80)(0.8)}{1214.7 - (80 - 32)} \times 100 = \frac{112}{1166.7} \times 100 = 9.6\%$

Answer is (C)

Evaporation:

M12:

By material balance,

$Fx_F = L\,x_L + V.y_v$

$y_v = 0$, F = 30000 lb/h, x_F = 0.1 mass fraction, x_L = 0.5 mass fraction

Substitution gives   30000(.1) = L(0.5)

L = 30000x0.1/0.5 = 6000 lb/h

Total evaporation = 30000 - 60000 = 24000 lb/h

Saturation temperature of water at 4"Hg= 1.96 psia = 125 °F

Enthalpy of superheated vapor = 1116 + 0.46(198 - 125) = 1149.6 Btu/lb

By heat balance, 30000 x 60 + S(1164 - 218.5) = 24000(1149.6) + 6000(222)

Solution of the above equation gives S = 28671 lb/h

Steam economy = 24000/28671 = **0.837** lb evaporation/lb of live steam

Answer is (C)

M13:

By heat balance on the condenser,
24000(1149.6) + W(54.03) = (W + 24000)(87.97)
From which,   W = 750048 lb/h
           = 750048/500 = 1500 gpm

Answer is (C)

Heat transfer in packed beds:
M14:

$$k_e = \frac{k_g}{D_t}\left(\frac{k_p}{k_g}\right)^{0.12}\left[3.4 + 0.00584\frac{D_p G}{\varepsilon \mu}\right]$$

$$\frac{k_g}{D_t} = \frac{0.0152}{0.2522} = 0.06028$$

$$\left(\frac{k_p}{k_g}\right)^{0.12} = \left(\frac{0.63}{0.0152}\right)^{0.12} = 1.5635$$

$$\frac{D_p G D}{\mu} = \frac{0.0015 \times 1000}{0.0435} = 34.48$$

$k_e = 0.06028 \times 1.5635 \times [3.4 + 0.00584 \times 34.48/0.42]$

    = 0.365   Btu/h.ft². °F/ft

Answer is (D)

Fouling:

M15:

Calculate first the log mean temperature difference.

$$\Delta T_1 = 450 - 310 = 140 \text{ °F} \quad \Delta T_2 = 350 - 300 = 50 \text{ °F}$$

$$\Delta T_{lm} = \frac{140 - 50}{\ln\frac{140}{50}} = 87.4 \text{ °F}$$

U_{do} = Dirty overall coefficient of heat transfer = $\frac{Q}{A \times \Delta T_{lm}}$

$$= \frac{490000}{200 \times 87.4} = 28.03 \text{ Btu/h.ft}^2.\text{°F}$$

Answer is (C)

M16:

Neglecting the metal wall resistance, relation for overall dirty coefficient of heat transfer can be written as follows

$$\frac{1}{U_{do}} = \frac{1}{h_o} + \frac{1}{h_{do}} + \frac{1}{h_i}\frac{d_o}{d_i} + \frac{1}{h_{di}}\frac{d_o}{d_i}$$

$$\frac{1}{28.03} = \frac{1}{38.4} + 0.003 + \frac{1}{300}\frac{1}{0.902} + \frac{1}{h_{di}}\frac{1}{0.902}$$

From which, $\frac{1}{h_{di}} = R_{di} = 0.00265$ h.ft².°F/Btu

Answer is (B)

KINETICS

Control of reactors:

M17:

Rewrite the given equation in the following form

$$\frac{V}{F + kV}\frac{dC_A}{dt} + C_A = \frac{F}{F + kV}C_{Ao}$$

In the above equation, coefficient, the coefficient of the derivative term is the time constant of the reactor and process gain K is the coefficient of C_{Ao}.

The equation is then written as $T\frac{dC_A}{dt} + C_A = KC_{Ao}$

Time constant for the reactor, $T = \frac{V}{F + kV} = \frac{V/F}{1 + kV/F} = \frac{1.6}{1 + 2(1.6)} = 0.381$ h

Answer is (B)

M18:

Step input is 0.72 - 0.75 = -0.03 moles/ft³

The transfer function with step input of -0.03 moles/ft³ is

$$C_1(s) = \frac{-0.03K}{s(Ts+1)}$$

$$K = \frac{F}{F+kV} = \frac{1}{1+kV/F} = \frac{1}{1+2(1.6)} = 0.238$$

Solution of the equation is $C_1(t) = -0.03(0.238)\{1 - e^{-t/T}\}$

At t = 2 h after imposition of change,

$$C_1(2\,h) = -0.00714\{1 - e^{-2/0.381}\} = -0.0071 \text{ mol/ft}^3$$

At steady state, $\frac{dC_A}{dt} = 0$ and from the transient material balance equation,

$$C_{As} = \frac{1}{1+kV/F} C_{Aos} = \frac{1}{1+2(1.6)}(0.75) = 0.1786 \text{ mols/ft}^3$$

Hence, $C_1(2\,h) = C_A - C_{as} = -0.0071$ mols/ft³

Therefore, outlet concentration after 2 hrs
$$= C_A(2h)$$
$$= 0.1786 - 0.0071 = 0.1715 \text{ mols/ft}^3$$

Answer is (C)

MASS AND ENERGY BALANCE

Conservation of energy:

M19:

Draw a sketch of the system as shown below.

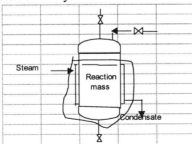

System : Take the system as shown by the envelop. Energy balance gives

$$U_E - U_B = \overset{(2)+(1)}{\Sigma(H+PE+KE)_I} - \overset{(2)+(1)}{\Sigma(H+PE+KE)_O} + \Sigma Q - \overset{3}{W}$$

1. No kinetic energy and potential energy changes
2. No mass in and out
3. No work involved

Energy balance reduces to

$$U_E - U_B = \Sigma Q = Q_s - Q_r - Q_l$$

For liquids little difference, if we replace $U_E - U_B$ by $H_E - H_B$

Therefore, $H_E - H_B = Q_s - Q_r -$

$H_E - H_B = \Delta H = Q_s - Q_r - Q_l$

$\Delta H = 500(0.8)(212 - 68) = 57600$ Btu/batch

Q_r = Heat of reaction = 991(500) = 495500 Btu/batch

Q_l = heat loss = 8000(2) = 16000 Btu/batch. in 2 hours

Therefore, Q_s = heat transferred to reaction mass from condensation of steam

$\qquad = 57600 + 495500 + 16000$

$\qquad = 569100$ Btu/batch

Steam usage, m = 569100/737.5 = **772 lb**

Answer (A)

Conservation of mass:

M20:

Basis = 100 mols of CH_4. Assume complete combustion.

$$CH_4 + 2O_2 = CO_2 + 2H_2O$$

	Feed mole		Products mole	O_2 used mole
CH_4	90	CO_2	90	90
N_2	10	H_2O	180	90
		N_2	10 + 0.79x	
		O_2	0.21x -180	

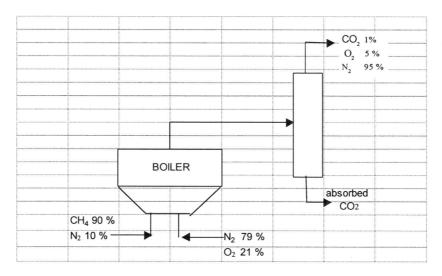

x = mols of air used.

Using N_2 as tie component,

$$\frac{oxygen\ in\ flue\ gas}{nitrogen\ in\ flue\ gas} = \frac{0.21x - 180}{10 + 0.79x} = \frac{5}{93.9}$$

Solving $19.719x - 16902 = 50 + 3.95x$ x = 1075 mols of air

N_2 = 10 + 1075(0.79) = 859.25 mols

O_2 = 1075x0.21 = 225.75 mols

O_2 in flue gas = $\frac{5}{93.9}(859.25) = 45.75$ mol

CO_2 = $\frac{1.1}{93.9}x(859.25) = 10.07$ mol

CO_2 = 90 - 10.07 = 79.93 mol

% CO_2 absorbed = $\frac{79.93}{90}(100) = 88.81$ %

Answer is (A)

M21:

Excess air : O_2 = 180 mols

O_2 in flue gas = 45.754 mols

Excess air = $\frac{45.754}{180}x100 = 25.42$ %

Answer is (C)

Economics:

M22:

$$\text{Annual cost} = (P - L)[A/P, i, n] + Li + OC$$

where P = initial cost, = $20,000
     L = salvage value, = $ 4,000
     A = annuity, $
     i = interest rate = 0.1
     OC = operating cost per year = $ 1000

*Capital recovery factor = $(A/P, i, n) = \dfrac{i(1+i)^n}{(1+i)^n - 1}$

*This factor can be calculated or obtained from the tables.

Annual cost = (20000 - 4000)[0.16275] + 4000(0.1) + 1000 = **$ 4,004**

Answer is (C)

Mass balance with and without reactions:
M23:

Overall balance : 100 = M + P

Where M = moles of NH_3 produced

P = purge, mols per hour

Overall H_2 balance :   100(0.751) = M(0) + P(0.8)

From equation second,   P = $\dfrac{100\,(0.751)}{0.8}$ = 93.875 moles/h

From first equation M = 100 - 93.875 = 6.125 moles/h

Now take hydrogen balance around mixing point of fresh feed and recycle

(0.751)100 + R(0.800) = (100 + R)(0.795)

75.1 + R(0.800) = (100 + R)(0.795)

From which, R(0.800 - 0.795) = 79.5 - 75.1 = 4.4

and R = 4.4/0.005 = **880 moles per hour**

Answer is (B)

M24:

Moles of hydrogen = 6.125(1.5) = 9.1875 moles/h

% conversion of hydrogen = $\frac{9.1875}{75.1} \times 100 = 12.23$ %

Answer is (A)

MASS TRANSFER

Forces and fluxes:

M25:

For diffusion of one component with second component non diffusing, the rate of diffusion of the diffusing component (In this example, oxygen) is given by

$$N_A = \frac{D_{AB}P}{RT z p_{BM}}(p_{A1} - p_{A2})$$

In this example, P = 1 atm.,

$p_{A1} = 0.13$ atm. $p_{B1} = 1 - 0.13 = 0.87$ atm.

$p_{A2} = 0.065$ atm. $p_{B2} = 1 - 0.065 = 0.935$ atm.

$p_{BM} = \frac{p_{B1} - p_{B2}}{\ln(p_{B1}/p_{B2})} = \frac{0.87 - 0.935}{\ln(0.87/0.935)} = 0.902$ atm.

Rate of diffusion of oxygen = $N_A = \frac{0.7224(1)}{0.7302(472)(0.12/12)(0.902)}(0.87 - 0.065)$

= 0.01515 lbmol/(h.ft²)

= 0.0303 lb/h.ft²

Answer is (A)

Liquid-solids systems:

M26:

Dry cake = 3.2 - 0.05(3.2) = 3.04 lb/ft³ of filtrate.

$\frac{wet\ cake}{Dry\ cake} = m = 3.2/3.04 = 1.053$

Mass fraction of solids in slurry = $x = \frac{3.04}{3.2 + 62.4} = 0.04634$

Since filter resistance is negligible, $V_1 = 0$, $\theta_1 = 0$

and the equation reduces to

$$V^2 = \frac{2A^2 \, \Delta P \, (1-mx)}{\mu \rho x a} \theta \qquad \text{and}$$

$$A^2 = \frac{V^2 \mu \rho x a}{2 (\Delta P)(1-mx)\theta} = \frac{60^2(1.1 \times 2.42)(62.4)(0.04634)(160)}{2(10 \times 144)(1 - 0.04634 \times 1.053)(40/60)} = 1877.22$$

A = 42.75 ft²

Answer is (A)

PVT data, equilibrium data:

M27:

$$T_c = 460 + (-180) = 278 \; °R \qquad P_c = 49.7 \text{ atm.}$$

To avoid trial and error calculation, determine V_r and T_r

Molar volume, $\bar{V} = \frac{1.04 \times 32}{8} = 4.16$ ft³/lb mol.

Molar critical volume, $\bar{V}_C = \frac{RT_C}{P_C} = \frac{10.73 \times 278}{49.7 \times 14.7} = 4.083$ ft³/lb mol

Reduced volume, $V_r = \frac{V}{V_C} = \frac{4.16}{4.083} \doteq 1.02$

Reduced temperature, $T_r = \frac{-13 + 460}{-180 + 460} = \frac{447}{280} \doteq 1.6$

From chart, at T_r = 1.6, and V_r = 1.02, P_r = 1.43

Then P = $P_r P_C$ = 1.43(49.7) = **71 atm.**

Answer is (C)

Separation systems:

M28:

[Using the given equilibrium diagram, the following points are located on it. The point D, composition of heptane in distillate. Since feed is saturated liquid, the q line is vertical and it is drawn from x_F = 0.4 to meet equilibrium curve at M. Joining D and M gives the operating line for minimum reflux. The construction is shown in the figure on page 29]

Minimum reflux operating line intersects y axis at y = 0.45

$$\frac{x_D}{R_M + 1} = 0.45 \qquad R_m = \frac{0.97}{0.45} - 1 = 1.156$$

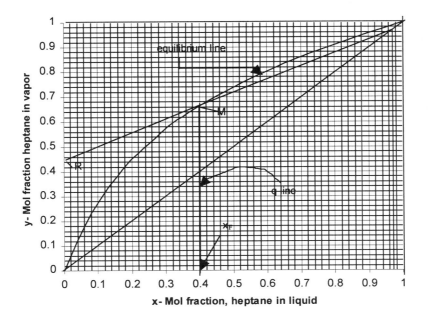

Answer is (B)

M29:

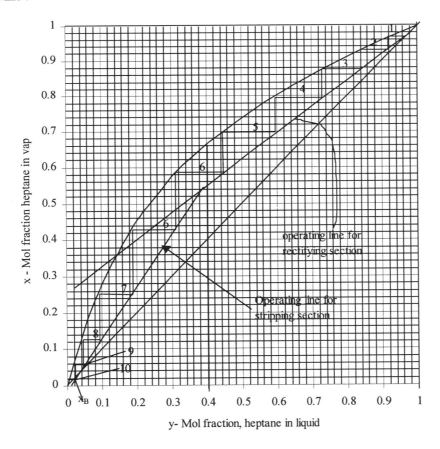

For reflux ratio of 2.5, the intercept on y axis = $\frac{x_D}{R+1} = \frac{0.97}{2.5+1} = 0.277$

A line is drawn from $x_D = 0.97 = y_D$ to intersect Y axis at y = 0.277
This is operating line for the rectifying section. It meets q line in Q.

A line is drawn from $x_B = y_B = 0.02$ to meet q line and rectifying section operating line in Q. This is operating line for stripping section. Theoretical stages are then stepped down as in the figure.

From the stage construction, number of stages = 10

Answer is (A)

Transport properties:

M30:

Ethyl acetate has the structure:

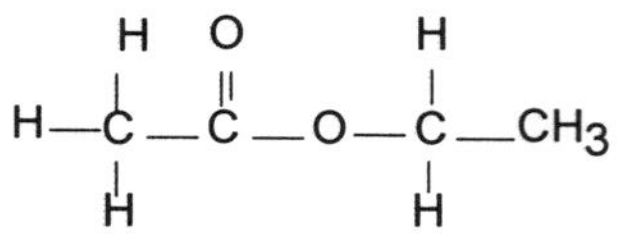

```
        H   O       H
        |   ||      |
    H — C — C — O — C — CH3
        |           |
        H           H
```

From the structure, C atoms = 4, H atoms = 8, O atoms = 2 and 1 Double bond

$\theta = 4(-0.462) + 8(0.249) + 2(0.054) + 1(0.478) = 0.73$

$T/T_C = \frac{273 + 20}{273 + 250.1} = 0.5601$

And $\log\left(\frac{8.569 \times \mu_L}{0.901^{0.5}}\right) = 0.73\left(\frac{1}{0.5601} - 1\right)$

$\mu_L = 3.74202 \times \frac{0.901^{0.5}}{8.569} = 0.415$ **cP = 0.415x2.42 = 1.004** $\frac{lb}{h.ft}$

Answer is (B)

Chemical properties:

M31:

2-methyl butane or isopentane undergoes combustion according to the following equation

$C_5H_{12} + 8O_2(g) = 5CO_2(g) + 6H_2O(l)$ $\Delta H_c = -843.216$ kcal/gmol

ΔH_f $\Delta H_f = 0$ 5 x (-94.052)   6 x (-68.316)

$\Delta H_R = \Delta H_c = \Sigma(\Delta H_f)_P - \Sigma(\Delta H_f)_R$

$$-843.216 = 5(-94.052) + 6(-68.316) - [\Delta H_f]_{isopentane}$$

From which, $(\Delta H_f)_{isopentane} = -36.93$ **kcal/gmol of C_5H_{12}**

Answer is (B)

Flooding and pressure drop:

M32:

Average molecular weight = $0.05(17) + 29(0.95) = 28.4$

$\rho_g = \frac{28.4}{359} \frac{492}{528} \frac{15}{14.7} = 0.0752$ lb/ft³

Gas flow rate = $30000(0.0752) = 2256$ lb/h

$\frac{L}{G}\sqrt{\frac{\rho_g}{\rho_L}} = 1\sqrt{\frac{0.0752}{62.4}} = 0.0347$ $\frac{\rho_w}{\rho_L} = \frac{62.4}{50} = 1.248$

For $\Delta P = 0.5$"H$_2$O, Y (ordinate) = 0.058 from pressure drop chart. (Perry's handbook)

$G_L^2 = \frac{0.0752(50)(4.18 \times 10^8)(0.058)}{52 \times 1.248 \times 0.23^2} = 1884640$, $G_L = 1373$ lb/h.ft²

$A = \frac{2256}{1373} = 1.643$ ft²

$D = \sqrt{\frac{1.643}{0.7854}} = 1.446 \doteq 1.5$ ft = 18"

Answer is (A)

M33:

At flood point and Abscissa = 0.0347, Y = 0.19

$\left(\frac{G_L}{G_f}\right)^2 = \frac{0.058}{0.19}$ $\frac{G_L}{G_f} = \sqrt{\frac{0.058}{0.19}} = 0.55$

Percent flood = $0.55(100) = 55\%$

Answer is (B)

Liquid-gas systems:

M34:

Mols of NH$_3$ in feed = 100 mols
Unrecvered NH$_3$ = 100 - 100(0.99) = 1 mol
Mol fraction, NH$_3$ in exit gas = 1/(900 + 1) = 0.00111

Maximum equilibrium concentration of NH_3 in liquid in equilibrium with feed composition = $x^* = 0.1/1.406 = 0.0711$.

$\left(\dfrac{L}{G}\right)_{min} = \dfrac{0.1 - 0.00111}{0.0711 - 0} = 1.391$

Actual ratio to be used = 2(1.39) = 2.78

Liquid rate = 2.78(1000) = 2780 lbmols/h = 2780(18) = 50040 lb/h

Cross section of tower = $0.7854(5.5)^2 = 23.76$ ft²

Rate of water = $\dfrac{50040}{500(23.76)} = 4.2$ gpm/h.ft²

Answer is (C)

M35:

Calculation of N_{OG}:

$y_1 = 0.1 \quad y_2 = 0.00111 \quad y_1^* = 0.07111 \quad y_2^* = 0$

$N_{OG} = \dfrac{y_1 - y_2}{(y - y^*)_{lm}}$

$(y - y^*)_{lm} = \dfrac{(0.1 - 0.07111) - (0.00111 - 0)}{\ln \dfrac{0.1 - 0.0711}{0.00111 - 0}} = 0.008524$

$\dfrac{y_1 - y_2}{(y - y^*)_{lm}} = \dfrac{0.1 - 0.00111}{0.008524} = 11.6$

G = 950.5/23.76 = 40 lbmol/h.ft²

$1 - y_1 = 0.9 \quad 1 - y_2 = 0.9989 \quad (1 - y)_{lm} = \dfrac{0.9989 - 0.9}{\ln \dfrac{0.9989}{0.9}} = 0.9486$

$H_{OG} = \dfrac{G}{K_Y a (1 - y)_{lm}} = \dfrac{40}{16.2(0.9486)} = 2.6$ ft

Height Z = $H_{OG}(N_{OG})$ = 11.6 (2.6) = 30.16 ft say 30.2 ft.

Answer is (B)

Phase diagrams:

M36:

Use **phase diagram** for $MgSO_4$. (Unit operations of chemical engineering, p-504 and p-858.)*

A horizontal line at 62 °F meets t vs concentration curve at 25 % concentration of mother liquor. Reading to the right end, the concentration of $MgSO_4 \cdot 7H_2O$ is 0.488.
Water in original solution = 65 lb per 100 lb solution.
Water evaporated = 0.02(65) = 1.3 lb per 100 lb of solution.
Water remaining = 65 - 1.3 = 63.7 lb.
New concentration of $MgSO_4$ = 35/(100 - 1.3) = 0.3546

By center of gravity principle,

Yield of $MgSO_4 \cdot 7H_2O$ = $\frac{0.3546 - 0.25}{0.488 - 0.25} \times (100 - 1.3)$

= 43.4 lb/100 lb of original sol.

Amount of crystals per ton of original solution = 43.4(2000/100) = **86 8 lb**

Answer is (B)

* (Note: In exam, the required diagrams will be provided with the questions)

Physical properties:

M37:

Van der Wall's equation is $\left(P + \frac{n^2 a}{V}\right)(V - b) = nRT$

P = 14.7 + 5.3 = 20 psia

Substituting the given values in the equation gives

$\left(20 + \frac{0.0044^2 \times 5.1 \times 10^3}{1.82}\right)(1.82 - 0.0044 \times 0.516) = 0.0044(10.73)T$

from which T = 772 °R = 772 - 460 = 312 °F

Answer is (B)

PLANT DESIGN

Control:

M38:

When a step change is applied to the system, the transform of y(t) becomes

$$Y(s) = \frac{1}{s} \frac{\tau_1 s + 1}{\tau_2 s + 1}$$

By final value theorem, $\lim_{t\to\infty} [y(t)] = \lim_{s\to 0} [sf(s)]$

$$= \lim_{s\to 0} s\left[\frac{1}{s}\frac{\tau_1 s + 1}{\tau_2 s + 1}\right]$$

$$= 1$$

By initial value theorem, $\lim_{t\to 0} [y(t)] = \lim_{s\to\infty} [sf(s)]$

$$= \lim_{s\to\infty} s\left[\frac{1}{s}\frac{\tau_1 s + 1}{\tau_2 s + 1}\right]$$

$$= \lim_{s\to\infty} \frac{\tau_1 + 1/s}{\tau_2 + 1/s}$$

$$= \frac{\tau_1}{\tau_2} = 5 \quad \text{(given)}$$

The value of y(t) varies betwee 5 and 1. The maximum value is therefore 5.

Answer is (B)

Equipment design:

M39:

$$\frac{L}{G}\sqrt{\frac{\rho_G}{\rho_L}} = \frac{121500}{79000}\sqrt{\frac{0.288}{46.4}} = 0.12$$

From Fig. 18-10, (Perry's Handbook p18-7)

for X = 0.12 and tray spacing of 24", Y ordinate = 0.32

$$U_f = 0.32 \left(\frac{\sigma}{20}\right)^{0.2}\left(\frac{\rho_L - \rho_g}{\rho_g}\right)^{0.5} = 0.32 \left(\frac{18}{20}\right)^{0.2}\left(\frac{46.4 - 0.288}{0.288}\right)^{0.5} = 3.97 \text{ ft/s}$$

Use U_a = 80 % of flood. U_a = 0.6(3.97) = 3.18 ft/s actual velocity.

Active area, $A_a = \frac{76.13}{3.18} = 23.94$ ft²

$A_t = A_a + 2A_d + A_w = 23.94 + 2(0.1)A_t + 0.075A_t$

Total area, $A_t = \frac{23.94}{0.725} = 33.02$ ft².

Diameter of column = $\sqrt{\frac{33.02}{0.7854}} = 6.484 \, ft \doteq 6.5$ ft

Answer is (C)

Thermodynamic laws:

M40:

The steady state energy balance reduces to

$$W = H_I - H_o - KE$$

By substituting given values,

$$W = 210.27 - 218.87 - \frac{200^2}{64.4(778)} = -9.4 \text{ Btu/lb}$$

Total work = $(9.4)(5)/42.4 = 1.1085 \doteq 1.11$ hp

Answer is (B)

END OF AM EXAM 1 SOLUTIONS..

SOLUTIONS TO AM EXAM 2
FLUIDS
Bernoulli's equation:

A1:

Applying Bernoulli equation to points A and B,

$$\frac{P_A}{\rho_A} + Z_A + \frac{U_A^2}{2g_c} - F + w = \frac{P_B}{\rho_B} + Z_B + \frac{U_B^2}{2g_c}$$

$Z_B = 0$ taken as datum plane. $w = 0$ no pump in line.

Bernoulli equation reduces to:

$$\frac{P_A}{\rho_A} + Z_A + \frac{U_A^2}{2g_c} - = \frac{P_B}{\rho_B} + \frac{U_B^2}{2g_c}$$

or $\quad \frac{P_A - P_B}{\rho_A} = F - Z_A - \left(\frac{U_A^2 - U_B^2}{2g_c}\right)$

$F = \frac{1.34}{12} \times \frac{1.05 \times 62.4}{0.0754} = 97.03$ ft of air.

$Z_A = 10$ ft (given) $\quad \frac{U_A^2 - U_B^2}{2g_c} = \frac{2.9^2 - 0.28^2}{64.4} = 0.13$ ft

$$\frac{P_A}{\rho_A} = \frac{P_o \times 144 + \frac{6}{12} \times 62.4 \times 2.5}{0.0754} = \frac{P_o \times 144}{0.0754} + 1034.5 \text{ ft of air}$$

$$\frac{P_B}{\rho_A} = \frac{P_o \times 144 + h_l \times 62.4 \times 1.05}{0.0754} = \frac{P_o \times 144}{0.0754} + \frac{h_l \times 62.4 \times 1.05}{0.0754} \text{ ft of air}$$

$$\frac{P_A - P_B}{\rho_A} = 1034.5 - \frac{h_l \times 62.4 \times 1.05}{0.0754} = 97.03 - 10 - 0.13$$

From which $h_l = 1.0905$ ft = 13.09"

Therefore, height from the bottom of tank = 13.09 + 3 = 16.1".

Answer is (B)

Control of flow systems:

A2:

At maximum flow ΔP_c is minimum and $\Delta P_c = 50$ psi

At minimum flow ΔP_c is maximum and $\Delta P_c = 110$ psi

C_{vc} at max flow = $1200 \sqrt{\frac{0.9}{50}} = 161$

C_{vc} at min flow = $100 \sqrt{\frac{0.9}{110}} = 9.0$

Required rangeability = 161/9 = 17.9 ≅ 18

Answer is (C)

Corrosion:

A3:

There is strongly acidic environment in the digester and the best material for the digestive sulfite process is brick lined vessel. In commercial operations, acid resistant brick lined vessels are used.

Answer is (A)

Economics:

A4:

Cost of replacement at the end of useful life = 25000 - 5000 = $ 20,000

$$\text{Capitalized cost} = K = C_v + \frac{C_R}{(1+i)^n - 1}$$

$$= 25000 + \frac{20000}{(1+i)^{10} - 1}$$

$$= 25000 + 17257.37 = \$\,42257.37$$

Answer is (A)

Flows in pipes and fittings:

A5:

Flow area of channel = 2x2 - 0.7854x0.84² = 3.446 in² = 0.02393 ft²

equivalent dia = $4r_H = \frac{4 \times 3.446}{[8 + \pi(0.84)]} = 1.2956" = 0.108$ ft = 1.3 in

$$u = \frac{100}{60 \times 7.48 \times 0.02393} = 9.31 \text{ ft/s}$$

$$\frac{Du\rho}{\mu} = \frac{0.108 \times 9.31^2 \times 62.4}{0.9 \times 0.000672} = 1.04 \times 10^5$$

f = 0.0236 from Moody chart at Re No. =1.04x10^5

$$\Delta P = \frac{0.0236 \times 50 \times 9.31^2 \times 62.4}{64.4 \times 0.108 \times 144} = 6.34 \quad psi$$

Answer is (B)

Sensors:

A6:

Assume turbulent flow. Therefore $C_o = 0.61$
Pitot tube positioned at the center gives the maximum velocity.

Density of air = $\frac{29}{359} \times \frac{492}{680} \times \frac{750}{760} = 0.05768$ lb/ft³

4" of water column = $\frac{4}{12} \times \frac{62.4}{0.05768} = 360.6$ ft of air

$u_{max} = 0.98\sqrt{64.4 \times 360.6} = 149.3$ ft/s

Average velocity u = 0.82(149.3) = 122.4 ft/s

Reynold's number = $\frac{D u \rho}{\mu} = \frac{(20/12)(122.4)(0.05768)}{0.022(0.000672)} = 1.064 \times 10^6$

Area of cross section = 0.7854(20/12)² = 2.182 ft²

Flow = 2.182(122.4)(60) = 16025 cfm at 750 mm and 220 °F

Therefore flow at 60 °F and 760 mm pressure,

$$= 16025(\tfrac{520}{680})(\tfrac{750}{760}) = 12093 \quad scfm$$

Answer is (B)

HEAT TRANSFER

Packed and fluidized beds:

A7:

For spheres $\phi_s = 1.0$. Assume $Re_p < 0.4$ since small particles

$$u_t = \frac{g(\rho_s - \rho_g) d_p^2}{18\mu} = \frac{981(2.5 - 1.2 \times 10^{-3})(0.005)^2}{18 \times 1.8 \times 10^{-4}} = 18.91 \quad cm/s$$

Check assumption of Re_p

$$Re_p = \frac{d_p u_t \rho_g}{\mu} = \frac{0.005(18.91)(1.2 \times 10^{-3})}{1.8 \times 10^{-4}} = 0.63$$

This is > 0.4 which was assumed. Therefore assume Re_p such that $0.4 < Re_p < 500$

Then $u_t = \left[\frac{4}{225} \frac{(\rho_s - \rho_g)^2 g^2}{\rho_g \mu}\right]^{\frac{1}{3}} d_p = \left[\frac{4}{225} \frac{(2.5 - 1.2 \times 10^{-3})^2 (981)^2}{(1.2 \times 10^{-3}) \times (1.8 \times 10^{-4})}\right]^{\frac{1}{3}} (0.005)$

= 40.3 cm/s

Check $Re_p = \frac{0.005 \times 40.3 \times 1.2 \times 10^{-3}}{1.8 \times 10^{-4}} = 1.34$ Thus $Re_p > 0.4$ and < 500

Therefore u_t = 40.3 cm/s is correct answer.

Answer is (C)

Resistance:

A8:

Inside surface area of the sphere = $\pi D_i^2 = \pi(4/12)^2 = 0.3461$ ft²

Outside surface area of the sphere = $\pi D_o^2 = \pi(8/12)^2 = 1.4103$ ft²

Mean area $A_m = \sqrt{A_i A_o} = 0.6986$ ft²

Heat loss, $q = \frac{k_m A_m \Delta t}{\Delta X} = \frac{26(0.6986)(300 - 220)}{2/12} = 8719$ Btu/h

Answer is (B)

Convection:

A9:

The expression for the Grashof number is given by

$N_{gr} = \frac{D_o^3 \rho_f^2 g \beta_f \Delta t}{\mu_f^2}$

$t_f = (220 + 100)/2 = 160$ °F $\Delta t = 220 - 100 = 120$ °F

$\rho_f = 61.28$ lb/ft³ $\mu_f = 1.221$ cP $g = 4.18 \times 10^8$ lb/h.ft

$\beta_f = \frac{0.0164 - 0.01624}{\frac{0.0164 + 0.01624}{2}(20)} = 4.9 \times 10^{-4}$ 1/°F

$N_{Gr} = \frac{0.0833^3 \times 61.28^2 \times 4.18 \times 10^8 \times 4.9 \times 10^{-4} \times 120}{(1.221 \times 2.42)^2} = 6.11 \times 10^6$

Answer is (B)

A10:

$$t_f = \frac{220 + 100}{2} = 160 \,°F$$

$\rho_f = 61.28$ lb/ft³ $\mu_f = 1.221$ cP $g = 4.18 \times 10^8$ lb/h.ft

$C_{pf} = 0.4751$ Btu/lb.°F $\beta_f = 4.9 \times 10^{-4}$ 1/°F

$k_f = 0.09379$ Btu/h.ft².°F/ft

$$h_c = 116 \left[\frac{k_f^3 \rho_f^2 C_{pf} \beta_f}{\mu_f'} \left(\frac{\Delta t}{d_o} \right) \right]^{0.25}$$

$$= 116 \left[\frac{0.09379^3 \times 61.28^2 \times 0.4751 \times 4.9 \times 10^{-4}}{1.221} \left(\frac{120}{1} \right) \right]^{0.25}$$

$$= \mathbf{59.9} \text{ Btu/h.ft}^2.°F$$

Answer is (A)

Energy conservation:

A11:

Examination of the data shows the following
(1) Blow down loss is already low. no much room for improvement.
(2) Flue gas temperature seems to be in normal range. Since its composition is not known, it is hard to say whether further heat recovery is possible.
(3) Since fuel is natural gas and not solid fuel, losses through ash are not possible.
(4) Excess air use is 90 % which is rather excessive. Normally for gaseous fuel, 10 to 15% excess air is used. Heat loss through excess air can be reduced.

Answer is (B)

Heat transfer in packed beds:

A12:

$$\frac{D_p}{D_t} = \frac{0.0015}{0.2522} = 0.00595 < 0.35$$

$$h_m = 0.813 \left(\frac{k_g}{D_t} \right) e^{-(D_p/D_t)} \left(\frac{D_p G}{\mu} \right)^{0.9}$$

$$= 0.813 \left(\frac{0.0152}{0.2522} \right) e^{-(0.0015/0.2522)} \left(\frac{0.0015 \times 1000}{0.0435} \right)^{0.9}$$

$$= \mathbf{1.144} \text{ Btu/h.ft}^2.°F$$

Answer is (B)

Radiation:

A13:

In this case, view factor = 1 and

Emissivity factor $F_e = \dfrac{1}{\frac{1}{\epsilon_1} + \frac{1}{\epsilon_2} - 1} = \dfrac{1}{1/0.9 + 1/0.75 - 1} = 0.6923$

$T_1 = 1200 + 460 = 1660$ R $T_2 = 1000 + 460 = 1460$ R

$\dfrac{Q}{A} = 0.6923(0.173)(16.8^4 - 14.6^4) = 3653$ Btu/h.ft²

Answer is (D)

A14:

OD of pipe = 2.38".
OD of pipe with insulation = 2.38 + 2 x 0.5 = 3.38"

Radiation area per foot of pipe = $\pi D_o L = \pi(3.38/12)(1) = 0.885$ ft²

$T_S = 125 + 460 = 585$ R $T_a = 70 + 460 = 530$ R

$q = A \in (0.173)\left[\left(\dfrac{T_S}{100}\right)^4 - \left(\dfrac{T_a}{100}\right)^4\right]$

$= 0.885(0.9)(0.173)\left[(5.85)^4 - (5.3)^4\right]$

= 52.6 Btu/h.ft².lin. ft of pipe

$h_r = \dfrac{q}{A(T_S - T_a)} \cong \dfrac{52.6}{0.885(585 - 530)} = 1.09$ Btu/h.ft².°F

Answer is (B)

Insulation:

A15:

log mean diameter = $\dfrac{4}{\ln\frac{5.315}{1.315}} = 2.864$ in

Insulation surface temperature, $t_s = 145$ °F (given)

$T_S = 460 + 145 = 605$ R $T_a = 460 + 70 = 530$ R

$\Delta t = 145 - 70 = 75$ °F

$$h_c = 0.5\left(\frac{\Delta t}{d_o}\right)^{0.25} \quad \text{for horizontal pipe}$$

$\quad = 0.5\left(\frac{75}{1.315}\right)^{0.25} = 1.37$ Btu/h.ft².°F

$h_r = 0.47(0.173)(6.05^4 - 5.3^4)/70 = 0.64$ Btu/h.ft².°F

$h_c + h_r = 1.37 + 0.64 = 2.01$ Btu/h.ft².°F

$q_s = 2.01 \times 75 = 150.8$ Btu/h.ft²

Surface area of insulation/ft of pipe $= \frac{5.315}{12} \times \pi \times 1 = 1.3915$ ft²

Heat loss per foot of insulated pipe $= 1.3915(150.8) = 209.8$ Btu/h

The reduction in heat loss $= 306 - 209.8 = 96.2$ Btu/h.ft² per ft of pipe

Answer is (C)

A16:

Critical radius of insulation $= \dfrac{k_{insulation}}{h_c + h_r} = \dfrac{0.12}{2.3} = 0.0522$ ft $= 0.63"$

Answer is (B)

KINETICS

Equilibrium: chemical/phase:

A17:

The equilibrium constant K at 25 °C is given by

$$\Delta G_f^0 = -RT \ln K$$

T = 273.15 + 25 = 298.15 K

Substituting the given data, $-3000 = -1.987 \, (298.15)(\ln K)$

or $\ln K = \dfrac{-3000}{-1.987 \times 298.15} = 5.064$

K = $e^{5.064}$ = **158.22**

If the heat of reaction is independent of temperature, the equilibrium constant K_T at a given temperature can be obtained by integration of Vant Hoff's relation

$$\ln \frac{K_T}{K_1} = \frac{-\Delta H_{rT}^0}{R}\left(\frac{1}{T_2} - \frac{1}{T_1}\right)$$

$T_2 = 273.15 + 50 = 323.15$ K $K_1 = 158.22$ at 25 °C

By substitution of given data

$$\ln \frac{K_T}{158.22} = \frac{-(-23990)}{1.987}\left(\frac{1}{323.15} - \frac{1}{298.15}\right)$$

$$\frac{K_{50}}{158.22} = e^{-3.132805} = 0.0436$$

$K_{50} = 0.0436(158.22) = 6.898392 \doteq 6.9$

Answer is (B)

A18:

For a first order reversible reaction with initial product concentration zero, it can be shown that

$$K_e = \frac{X_{AE}}{1 - X_{AE}} \quad \text{or} \quad X_{AE} = \frac{K_E}{K_E + 1}$$

At t = 50 °C, K = K_E = 6.9 as calculated in problem KC-1

$$X_{AE} = \frac{6.9}{6.9 + 1} = \frac{6.9}{7.9} = 0.8734$$

$$= 87.34 \%$$

Answer is (A)

MASS AND ENERGY BALANCE

Conservation of energy:

A19:

Assume 77 °F as datum temperature.

H of $CaCO_3$ at 77 °F = 0
H of $CaCO_3$ at 400 °F = (400 - 77) x 0.242 = 78.17 Btu/lb

H of CO_2 at 1832 °F = (1832 - 77)(0.2830) = 496.7 Btu/lb
H of CaO at 1832 °F = (1832 - 77)(0.245) = 430 Btu/lb

By energy balance, Q to be supplied

= Heat of reaction + Enthalpy of products - Enthalpy of reactants

= 20000(788.4) + 8800(496.7) + 11200(430.7) - 20000(78.17)

= 23.4 x 10^6 Btu/h

With kiln efficiency of 80 %, heat to be supplied
= 23.4x10^6/0.8 = 29.25x10^6 Btu/h

= **29.25 MMBtu/h**

Answer is (B)

Conversion/yield:

A20:

$$CaCO_3 \rightarrow CaO + CO_2$$
$$100 \quad\quad 56 \quad\; 44$$

Basis : 100 lb of $CaCO_3$
Reaction is 95 % complete.
$CaCO_3$ decomposed = 95 lb per 100 lb $CaCO_3$

CO_2 produced = $\frac{44}{100} \times 95 = 41.8$ lb

Therefore, yield of CO_2 = 41.8/100 = 0.418 lb per lb limestone.

Answer is (B)

A21:

$$Al_2O_3 + 3H_2SO_4 \rightarrow Al_2(SO_4)_3 + 3H_2O$$
$$101.9 \quad\; 3 \times 98 \quad\quad 342.1 \quad\;\; 3 \times 18$$

Al_2O_3 in Bauxite feed = 0.554(2160) = 1196.64 lb

$Al_2(SO_4)_3$ produced = 3600 lb

$$Al_2O_3 \text{ used} = \frac{101.9}{342.1} \times 3600 = 1072.3 \text{ lb}$$

$$\% \ Al_2O_3 \text{ used} = \frac{1072.3}{1196.64} \times 100 = 89.61 \ \%$$

Answer is (C)

Economics:

A22:

$$\text{Capitalized cost} = P + \frac{A}{i} + \frac{(P - L)(A/F, i, n)}{i}$$

In this case, P = $300,000 initial investment
        A = perpetual annual cost or perpetual equivalent annual cost of a future one time cost
        L = salvage value.
        (A/F, i, n) = sinking fund factor

Since the asset life is too long, the third term can be ignored.

A = 1000 + 7000(A/F, i, n) = 1000 + 7000(0.06903) = $1483.21

$$\text{Therefore, capitalized cost} = 300000 + \frac{1483.21}{0.08} = \$ \ 318540.13$$

Answer is (B)

MASS TRANSFER

Cooling towers:

A23:

$$H_1 = 38.6 \quad \text{Btu/lb}$$

$$H_2 = H_1 + \frac{LC_L}{G}(t_{l2} - t_{L1}) = 38.6 + \frac{1302 \times 1}{1400}(120 - 85) = 71.15 \text{ Btu/lb}$$

The operating line is drawn with these points and corresponding temperatures.

$$(NTU)_G = \int \frac{dH}{H_i - H}$$

To estmate the integral quickly, use Simpson's rule.
Estimate slope enthalpy interface lines as follows

$$-\frac{h_l a}{k_g a M_B P} = \frac{0.03 \ (1302)^{0.51}(1400)}{0.04 \ (1302)^{0.26}(1400)^{0.72}(29)(1)} = -1.18$$

With this slope draw three lines to intersect the operating line and saturation enthalpy curve as shown in the following figure.

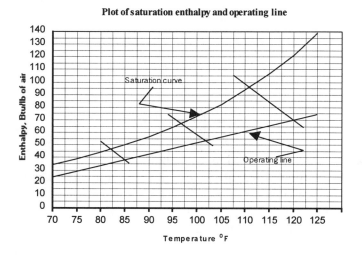

Read H_i and H from the intersection lines as follows

H_i	H	$H_i - H$	$\frac{1}{H_i - H}$
47	38.6	8.4	0.119
68.2	54.9	13.3	0.0752
96	71.15	24.85	0.0402

By Simpson's rule,

$$\int \frac{dH}{H_i - H} = \frac{\frac{71.15 - 38.6}{2}}{3}[0.11905 + 4(0.0752 + 0.0402)]$$

$$= 2.496 \doteq 2.5$$

$$(NTU)_G = 2.5$$

Answer is (A)

A24:

From humidity chart, for air with 85 °F dry bulb and 75 °F wet bulb temperature, the humidity is near to 0.0168 lb of water vapor per lb/dry air.

Answer is (C)

Forces and fluxes:

A25:

$$\Gamma = 125 \ \ lb/h.ft \quad D_{AB} = 7.6 \times 10^{-5} \ ft^2/h \quad \rho = 62.3 \ lb/ft^3 \quad \delta = 7.94 \times 10^{-4} \ ft$$

Velocity in y direction $\bar{U}_y = \frac{\Gamma}{\rho\delta} = \frac{125}{62.3 \times 7.94 \times 10^{-4}} = 2527 \, ft/h = 0.702 \, ft/s.$

At top of the column, $C_{AO} = 0$, At bottomm of column = $C_{AO} = C_{Ab}$

$C_{Ai} = 0.002$ lbmol/ft³

Therefore, at top of column, driving force = $C_{Ai} - C_{At} = 0.002 - 0 = 0.002$ lbmol/ft³

At bottom of column, the driving force = $(C_{Ai} - C_{Ab})$

$$N_{A,av} = \frac{\bar{U}_y \delta}{L}(C_{A,L} - C_{A0}) = K_{L,av}(C_{Ai} - C_A)_{mean}$$

Where $(C_{Ai} - C_A)_{mean} = \frac{[(0.002 - C_{At}) - (0.002 - C_{Ab})]}{\ln\left(\frac{0.002}{0.002 - C_{Ab}}\right)}$

Equating the two expressions for flux,

$$N_{A,av} = \frac{2527 \times 7.94 \times 10^{-5}}{4} C_{Ab} = 0.3028 \frac{[0.002 - (0.002 - C_{Ab})]}{\ln\left(\frac{0.002}{0.002 - C_{Ab}}\right)}$$

From which $C_{ab} = 0.0009063$ lbmol/ft³
And the rate of absorption is $\bar{U}_y \delta (C_{Ab} - C_{At})$

$$= 2527(7.94 \times 10^{-4})(0.0009063 - 0)$$

$$= 1.818 \times 10^{-3} \text{ lbmol/h.ft of film width.}$$

Answer is (B)

Gas-solid systems:

A26:

Make enthalpy and moisture balances. The operation can be represented as follows

$H_1 = 0.24(110) + Y_1[1061 + 0.45(110)] = 26.4 + 1110.5 Y_1$ Btu/lb dry air

Enthalpy balance on the dryer gives (assume M = mass of dry air lb/h)

$71.9M + 14.4 \times 1700 + 48.02 \times 300 = M(26.4 + 1110.5 Y_1) + 8 \times 107.88 + 1700(32.4)$

which simplifies to $M(45.5 - 1110.5 Y_1) = 17057.04$

Moisture balance : $M(Y_1 - 0.01) = 1700(0.1765 - 0.005)$

Dividing one equation into the other

$$\frac{45.5 - 1110.5 Y_1}{Y_1 - 0.01} = \frac{17057.04}{1700 \times 0.1705} = 58.505$$

$$Y_1 = \frac{45.5 + 0.5805}{1164.505} = 0.0396 \text{ lb/lb dry air}$$

$$M = \frac{1700 \times 0.1715}{0.0396 - 0.01} = 9850 \text{ lb/h dry air}$$

Answer is (C)

A27:

$$\text{Cross section of dryer} = \frac{\pi (5)^2}{4} = \frac{\pi \times 25}{4} = 19.635 \, ft^2,$$

$G = 9850/19.635 = 502$ lb/h.ft²

$\Delta t_1 = 110 - 80 = 30$ °F $\Delta t_2 = 240 - 140 = 100$ °F

$$\Delta t_{lm} = \frac{100 - 30}{\ln \frac{100}{30}} = 58.14 \text{ °F}$$

$$N_t = \frac{\Delta t_G}{\Delta t_{lM}} = \frac{240 - 110}{58.14} = 2.236 \quad \text{heat transfer units}$$

Answer is (B)

Liquid-solid systems:

A28:

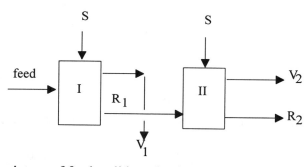

Total mixture of fresh solids and solvent = 3000 + 1000 = 4000 lb/h

Mass fraction solvent solvent in mixture = 3000/4000 = 0.75

Mass fraction solute in mixture = 500/4000 = 0.125

$R_1 = 500 + 500(1.2) = 1100$ lb/h

$E_1 = 4000 - 1100 = 2900$ lb/h

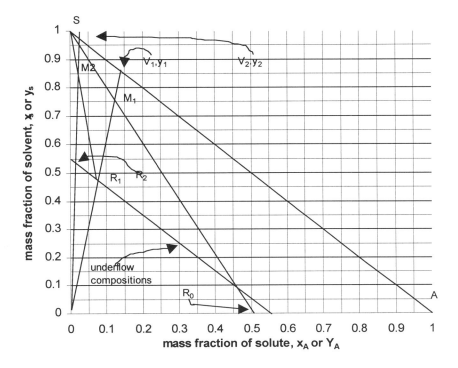

See the figure above for solution

Locate R_0 on X axis at $x_A = 0.5$. Join R_0 and S and locate Point M1 representing the mixture with coordinates $x_s = 0.75$ and $x_A = 0.125$. Join M1 and origin to intersect the underflow curve in R_1 and the hypotenuse in y_i, V_1.

From diagram, $y_1 = 0.86$, $R_1 + S = 1100 + 3000 = 4100 = M2$

Solvent in M2 = $3000 + 600(0.86) = 3516$

Join SR_1. Locate M2 at $x_s = 0.86$. Join origin and M2 to meet hypotenuse in V_2, y_2

$y_2 = 0.98$ $V_2 = 4100 - 1100 = 3000$ lb/h

Solute recovered in 2 stages = $2900(0.14) + 3000(0.02) = 466$ lb/h

Recovery = $\frac{466}{500} \times 100 = 93.2$ %

Answer is (D)

PVT data, equilibrium data:

A29:

[Assume Raoult's law applies to water in saturated water phase and to ether in saturated ether phase. In water phase, Henry's law applies to ether. then

$$y_B P = x_B k_B = (1 - x_A) k_B \quad \text{by Henry's law}$$

and $\quad y_A P = x_A P_A \quad$ by Raoult's law

adding the two gives $\quad (y_A + y_B) P = (1 - x_A) k_B + x_A P_A = P$ since $y_A + y_B = 1.0$

$x_A = \frac{P - k_B}{P_A - k_B} \quad$ and $\quad y_A = \frac{P_A x_A}{P}$]

Using these relations and the given data, first calculate k_B at 50 °C

Thus, $x_A = 1 - x_B = 1 - 0.0103 = \frac{1.744 - k_B}{0.121 - k_B}$

From which, $\quad k_B = 168$ at 50°C.

Now at 1 atm., $x_A = 1 - x_B = \frac{1 - 168}{0.121 - 168} = 0.9948$

and $x_B = 0.0052$ mol fraction ether

$$y_A = 1 - y_B = \frac{P_A x_A}{P} = \frac{0.121(0.9948)}{1} = 0.1204$$

and $y_B = 1 - 0.1204 = 0.8796 \doteq 0.88$

Answer is (B)

Transport properties:

A30:

$$T = 273.1 + 80 = 353.1 \text{ K}$$

<u>Total diffusion volume for methyl chloride</u>

		Air
1 carbon	= 16.5	20.1
3 hydrogen (3x1.98)	= 5.94	
1 chlorine	= 19.5	
	Total = 41.94	

$M_A = 50.5 \quad M_B = 29 \quad P = 1$ atm

$$D_{AB} = \frac{10^{-3} \times 353.1^{1.75}\left[\frac{1}{50.5} + \frac{1}{29}\right]^{0.5}}{(1)\left[(41.94)^{1/3} + (20.1)^{1/3}\right]^2} = 0.175 \text{ cm}^2/\text{s}$$

Answer is (A)

Chemical properties:

A31:

First find ΔH_o using T = 298 K where heat of reaction is available.

$$\Delta H_0 = -39.43 + 10^{-3}\left\{-\left[\Delta a T + \frac{\Delta \beta}{2}T^2 + \frac{\Delta \gamma}{3}T^3\right]\right\}$$

Where 10^{-3} factor converts cal to kcal

$$= -39.93 + 10^{-3}\left\{-\left[-14.945 \times 298 + \frac{11.082 \times 10^{-3}}{2} \times 298^2 - \frac{1.729 \times 10^{-6}}{3} \times 298^3\right]\right\}$$

$= -39.43 + 4.454 - 0.492 + 0.0153 =$ **-35.55 kcal/gmol**

Heat of reaction at 500°C = 773 K is given by

$$(\Delta H_R)_{773} = -35.55 - 14.945 \times 773 + \frac{11.02 \times 10^{-3}}{2} \times 773^2 - \frac{1.729 \times 10^{-6}}{3} \times 773^3$$

= - 35.55 - 11.553 + 3.31 - 0.266 = **-43.964 kcal/gmol**

Answer is (B)

Condensation of mixed fluids:

A32:

The dew point of the mixture from the given data of condensation is 213.8 °F.

Similarly the bubble point is 185 °F. The condensing range is therefore 213.8-185 °F.

Heat load in condensing range = 13.9792 - 1.2874 = 12.1518 MM Btu/h

Answer is (A)

A33:

Divide the entire cooling range into several small intervals nd calculate Δt for each interval. Use arithmetic averages. The following table is prepared.

hot fluid temp °F	cold fluid temp °F	Δt °F	Δt_{av} °F*
230	120	110	
220	113.46	106.54	108.27
213.8	101.26	112.54	109.54
210	87.73	122.17	117.41
190	81.22	108.78	113.4
185	80	105.	106.89

*As an example, in first interval, $\Delta t_1 = 110 \quad \Delta t_2 = 106.54$

$$\text{And } \Delta t_{av} = \frac{110 + 106.54}{2} = 108.27 \text{ °F}$$

Similar calculations are made for other intervals

Weighted average

$$\Delta t = \frac{0.39 \times 108.27 + 0.241 \times 109.54 + 1.8396 \times 117.41 + 4.3248 \times 113.4 + 2.0906 \times 106.89}{12.7833}$$
$$= 115 \text{ °F}$$

Answer is (B)

Humidification:

A34:

Lewis number is the ratio of Schmidt number to Prandtl number.

Prantdl number for nitrogen = $\frac{C_P \mu}{k} = \frac{0.282 \times 0.0127 \times 2.42}{0.0115} = 0.754$

$\rho = 0.0925$ lb/ft³

Schmidt number = $\frac{\mu}{\rho D_{AB}} = \frac{0.0127 \times 2.42}{0.0925 \times 0.523} = 0.635$

Lewis number = $\frac{Sc}{Pr} = \frac{0.635}{0.754} = 0.842$

Answer is (B)

A35:

Assume first that nitrogen is saturated at wet bulb temperature.

$p_A = 255$ mm Hg $Y" = \frac{255}{800-255} \times \frac{58}{28} = 0.97$ lb/lb nitrogen

$C_S = 0.248 + 0.97(0.318) = 0.53$ Btu/lb dry air

$\frac{k_G}{k_y} = 0.53 \, (Le)^{0.567} = 0.53(0.842)^{0.567} = 0.48$

$t_G - t_w = \frac{\lambda_w(Y'_w - Y')}{k_G/k_y}$ wet bulb equation

$24 = \frac{235(0.97 - Y')}{0.48}$ $Y' = 0.97 - \frac{24 \times 0.48}{235} = 0.921$

$C_S = 0.248 + 0.921(0.318) = 0.54$

$\frac{k_G}{k_y} = 0.54 \, (Le)^{0.567} = 0.54(0.842)^{0.567} = 0.49$

$24 = \frac{235(0.97 - Y')}{0.49}$ $Y' = 0.97 - \frac{24 \times 0.49}{235} = \mathbf{0.920}$ lb/lb dry air o.k.

Answer is (A)

A36:

Use the given phase diagram for solving tha problem.

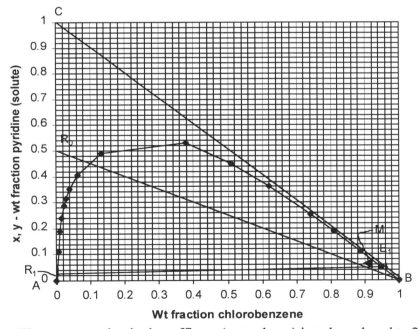

Pyridine-water-chlorobenzene system

The concentration in the raffinate (water layer) is to be reduced to 2 %. This point must lie on the distribution curve and at y = 0.02. Utilizing this fact, the construction is made as shown in the figure above.

The point R_0 (original pyridine-water solution) is connected with point B representing the solvent. Point R_1 is located on the distribution curve and the tie line passing through it is drawn to intersect R_0B in M, the mixture point and to meet the distribution curve in E_1. The amount of solvent used is found from line segment lengths R_0M and MB.

Solvent used = $\frac{R_0M}{MB} = \frac{82}{7.5} \times 2000 = 21867$ lb/batch.

Answer is (B)

Physical properties:

A37:

$$\Delta H_v = T\Delta V \frac{dP}{dT}$$

For approximate solution, we can write the equation as

$$\Delta H_V = T\Delta V \frac{\Delta P}{\Delta T}$$

Now $\frac{\Delta P}{\Delta T} = \frac{32.624 - 27.134}{290 - 270} = 0.2745$ atm/°R

Therefore, $\Delta H_v = (460 + 280)(0.165 - 0.0429)(0.2745)$ °R$\left(\frac{ft^3}{lb}\right)\left(\frac{atm.}{°R}\right)$

$= 24.80$ °R$\left(\frac{ft^3}{lb}\right)\left(\frac{atm}{°R}\right) \times \left(144 \times 14.7 \frac{lb}{ft^2.atm}\right)$

$= \frac{52496.64}{778} \frac{\frac{ft.lb}{lb}}{\frac{ft.lb}{Btu}} = 67.5$ Btu/lb

Answer is (C)

Economics:

M38:

Product produced per year = 28200 (360)/2000 = 5076 tons

Turn over ratio = $\frac{annual\ sales\ \$}{fixed\ capital\ \$} = 1$ **given**

Therefore, annual sales = Fixed capital = 1500000 - 180000 = $1320000

If C is cost/ton in dollars,
5076 x C = 1320000 which gives C = 1320000/5076 = **260.05 $/ton**

Answer is (D)

A39:

$$P = F \left(\frac{1}{1+i}\right)^n \quad i = 5\% \quad n = 4 \text{ years remaining}$$

$$2000 = 3000 \left(\frac{1}{1+i}\right)^4 \quad \text{or} \quad 1+i = \left(\frac{3000}{2000}\right)^{\frac{1}{4}} = 1.1067$$

Therefore, $i = 1.1067 - 1 = 0.1067$ or **10.67 %**

Answer is (D)

Control:

A40:

The characteristic equation of the control system can be written as

$$1 + \frac{K_C}{(s+1)(0.5s+1)} \times \frac{3}{s+3} = 0$$

This is simplified to $(s+1)(0.5s+1)(s+3) + 3K_C = 0$

Further simplification gives $\frac{1}{6}s^3 + s^2 + \frac{11}{6}s + 3(1+K_C) = 0$

We prepare Ruth's array as follows

1	$\frac{1}{6}$	$\frac{11}{6}$
2	1	$1 + K_C$
3	$\frac{10 - K_C}{6}$	
4	$1 + K_C$	

In order to have all elements of first row positive, $K_C < 10$ since K_C is positive. The system will be stable if K_C is less than 10. However, Ruth's test is not adequate to tell about the degree of stability or oscillatory nature of the system. For this, other tests are needed.

Answer is (A)

END OF SOLUTIONS TO AM EXAM 2.

AFTERNOON SAMPLE EXAMINATION

Instructions for afternoon Session

1. You have four hours to work on the afternoon session. Do not write in this handbook.

2. Answer all forty questions for a total of forty answers. There is no penalty for guessing.

3. Work rapidly and use your time effectively. If you do not know the correct answer, skip it and return to it later.

4. Some problems are presented in both metric and English units. Solve either problem.

5. Mark your answer sheet carefully. Fill in the answer space completely. No marks on the workbook will be evaluated. Multiple answers receive no credit. If you make a mistake, erase completely.

Work all 40 problems in four hours.

P.E. Chemical Engineering Exam
Afternoon Session

Ⓐ Ⓑ Ⓒ Fill in the circle that matches your exam booklet.

MA1 Ⓐ Ⓑ Ⓒ Ⓓ	MA11 Ⓐ Ⓑ Ⓒ Ⓓ	MA21 Ⓐ Ⓑ Ⓒ Ⓓ	MA31 Ⓐ Ⓑ Ⓒ Ⓓ
MA2 Ⓐ Ⓑ Ⓒ Ⓓ	MA12 Ⓐ Ⓑ Ⓒ Ⓓ	MA22 Ⓐ Ⓑ Ⓒ Ⓓ	MA32 Ⓐ Ⓑ Ⓒ Ⓓ
MA3 Ⓐ Ⓑ Ⓒ Ⓓ	MA13 Ⓐ Ⓑ Ⓒ Ⓓ	MA23 Ⓐ Ⓑ Ⓒ Ⓓ	MA33 Ⓐ Ⓑ Ⓒ Ⓓ
MA4 Ⓐ Ⓑ Ⓒ Ⓓ	MA14 Ⓐ Ⓑ Ⓒ Ⓓ	MA24 Ⓐ Ⓑ Ⓒ Ⓓ	MA34 Ⓐ Ⓑ Ⓒ Ⓓ
MA5 Ⓐ Ⓑ Ⓒ Ⓓ	MA15 Ⓐ Ⓑ Ⓒ Ⓓ	MA25 Ⓐ Ⓑ Ⓒ Ⓓ	MA35 Ⓐ Ⓑ Ⓒ Ⓓ
MA6 Ⓐ Ⓑ Ⓒ Ⓓ	MA16 Ⓐ Ⓑ Ⓒ Ⓓ	MA26 Ⓐ Ⓑ Ⓒ Ⓓ	MA36 Ⓐ Ⓑ Ⓒ Ⓓ
MA7 Ⓐ Ⓑ Ⓒ Ⓓ	MA17 Ⓐ Ⓑ Ⓒ Ⓓ	MA27 Ⓐ Ⓑ Ⓒ Ⓓ	MA37 Ⓐ Ⓑ Ⓒ Ⓓ
MA8 Ⓐ Ⓑ Ⓒ Ⓓ	MA18 Ⓐ Ⓑ Ⓒ Ⓓ	MA28 Ⓐ Ⓑ Ⓒ Ⓓ	MA38 Ⓐ Ⓑ Ⓒ Ⓓ
MA9 Ⓐ Ⓑ Ⓒ Ⓓ	MA19 Ⓐ Ⓑ Ⓒ Ⓓ	MA29 Ⓐ Ⓑ Ⓒ Ⓓ	MA39 Ⓐ Ⓑ Ⓒ Ⓓ
MA10 Ⓐ Ⓑ Ⓒ Ⓓ	MA20 Ⓐ Ⓑ Ⓒ Ⓓ	MA30 Ⓐ Ⓑ Ⓒ Ⓓ	MA40 Ⓐ Ⓑ Ⓒ Ⓓ

PM SAMPLE EXAM 1
FLUIDS

Bernoulli's equation:

MA1: In the figure below is shown a tank discharging water to atmospheric pressure through a 2" standard pipe. Total frictional and contraction losses in exit line amount to 28.5 feet of water head. With water level held constant in the tank, the discharge flow is 101.2 gpm. Density of water is 62 lb/ft³.

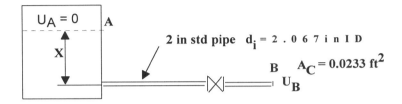

Under these conditions the water level [ft] from the centerline of the pipe is closest to:

(A) 36   (B) 26   (C) 20   (D) 30

MA2: A centrifugal pump draws a solution of specific gravity 1.605 g/ml from a storage tank of large cross section through a std. 3" schedule 40 pipe and delivers it to an overhead tank . (see sketch below). Liquid velocity in the suction line is 3 ft/s. The pump discharges through a 2" schedule 40 std. pipe The level of solution in the tank is maintained at 10 ft from the centerline of the pump. The discharge point is 40 ft above the centerline of the pump but the liquid level in

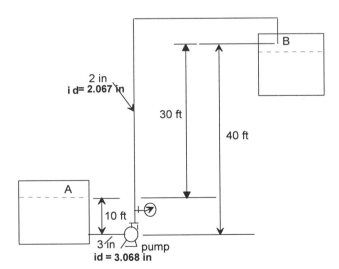

the overhead tank is always below the discharge point. The losses due to friction and contraction in the suction and discharge sections of the piping are equivalent to 12 ft of head of solution.

The pressure [psig] indicated by the gauge at the pump discharge is close to:

(A) 29.7 (B) 12 (C) 9.0 (D) 15

MA3: Water (density = 62.4 lb/ft³, viscocity = 1 cP) is pumped from a large reservoir on the floor at a rate of 80 gpm to the top of a small tower. The discharge point is 20 ft above the centerline of the pump. Frictional and other (contraction, expansion etc.) losses in suction and discharge lines amount to 1.5 ft head of liquid. If the pump develops only ¼ HP, the water height [ft] in the reservoir must be closest to:

(A) 20 (B) 10 (C) 13 (D) 15

MA4: An oil with a density of 56.1 lb/ft³ enters a piping system at 20 psia, and a velocity of 15 ft/s. It discharges downstream at a point which is 20 ft below the inlet. The velocity at the outlet is 10 ft/s. The fluid gains 0.005 Btu/lb as heat during its passage through the piping system. at a mass flow rate of 40 lb/s. Under these conditions, the pressure (psia) at the outlet is near to

(A) 25 (B) 30 (C) 15.3 (D) 20.6

Corrosion:

MA5 to MA7: Hydrolysis of a chlorophosporus organic waste residue from a plant manufacturing insecticides is to be carried out in vessels of suitable material of construction. The hydrolysis will take place at a pressure slightly less than atmospheric (1 to 2 in of water column below atmospheric). The temperature in the hydrolysis vessel is to be maintained at between 80 and 90 °C. The residue is added to the hydrolyzer vessel at controlled rate so that peak rate of gas evolution does not exceed design rate for pollution abatement facility. The products of hydrolysis are

        Gaseous   : HCl, H_2S, CH_3OH, C_2H_5OH, water vapor
        In solution : H_3PO_4, H_2SO_3, H_2SO_4, HCl

MA5: As a process engineer, which of the following materials of construction you would recommend for the hydrolyzer?

    (A) Brick lined vessel
    (B) Haveg
    (C) Hastelloy C
    (D) Rubber lined steel

MA6: Heat to the hydrolyzer is to be supplied by an immersion heater of suitable material of construction. The heater uses steam at 75 psig. The suitable material of construction for this heater is

    (A) Stainless steel 316
    (B) Karbate
    (C) Teflon
    (D) Hastelloy B

MA7: The gases from the hydrolyzer in problem MA5 are combusted for the purpose of further treatment to meet pollution abatement standards. Products of combustion, viz SO_2, SO_3, HCl, CO_2, water vapor, and N_2, O_2 leave the combustion chamber at about 1800 °F and are to be quenched in a tower type vessel. A suitable material for the lining of the steel vessel is

        (A) Acid-resistant brick
        (B) Glass
        (C) Monel
        (D) Rubber

MA8: Of the following four methods, the one that will actually increase the galvanic corrosion of more active metal is

        (A) Using combination of two metals as close as possible in galvanic series.
        (B) Coupling two widely separated metals in the galvanic series
        (C) Protective oxide films
        (D) Insulating the two metals from each other

Flow in pipes and fittings:

MA9 to MA12:

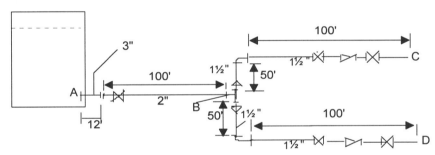

As shown in the sketch above, water at 70° F is discharged from a tank to two points C and D. Flow through each 1½" line is 20 GPM. Points A, B, C, and D are at the same elevation. Additional data are as follows:

    density of water = 62.3 lb/ft³   viscosity of water = 0.982 cP

Dia. (nom)	id in	id ft	A_c ft²	u ft/s	ϵ/d_i	Re No.	f
3"	3.068	0.2557	0.05134	1.74	0.00059	4.2x10⁴	0.024
2"	2.067	0.1753	0.02330	3.83	0.00087	6.23x10⁴	0.023
1½"	1.61	0.1342	0.01415	3.15	0.00112	3.99x10⁴	0.024

The following nomenclature is defined for the section AB of the piping:

    ΣK = Sum of the velocity constants representing head loss due to fittings, contraction, and expansion
    L = Length of piping in section AB, based on 2" schedule 40 pipe

d_i = inside diameter in
f = Moody friction factor
u = velocity in ft/s
h_f = head loss due to friction in ft of liquid.

MA9: The correct relationship to be used to calculate h_f for section AB of the piping is:

(A) $h_f = \dfrac{4fL}{d_i} \dfrac{u^2}{2g_c}$

(B) $h_f = \left(\Sigma K + \dfrac{fL}{d_i}\right) \dfrac{u^2}{2g_c}$

(C) $h_f = \left(\Sigma K + \dfrac{4fL}{d_i}\right) \dfrac{u^2}{2g_c}$

(D) $h_f = \dfrac{fL}{d_i} \dfrac{U^2}{2g_c}$

MA10: Moody friction factor for flow in 2" pipe is most nearly:

(A) 0.0058 (B) 0.023 (C) 0.019 (D) 0.031

MA11: For the projected entrance of 3" schedule 40 pipe in the tank, K = 0.78. The K value based on 2" schedule 40 pipe will be most nearly:

(A) 0.78 (B) 0.53 (C) 1.0 (D) 0.16

MA12: The Reynolds number for the 1½" schedule 40 pipe is most nearly:

(A) 1.1×10^5 (B) 8.0×10^4 (C) 3.99×10^4 (D) 2×10^4

Packed and fluidized beds:

MA13 to MA16: Solids of the following size distribution are to be fluidized:

Cumulative weight of representative sample	With dia. < d_p ft
0.00	0.000164
0.166	0.000246
0.417	0.000328
0.749	0.000410
0.917	0.000492
1.00	0.000574

The calculated mean diameter of the particles is 0.322×10^{-3} ft. The other data for the solids and the fluidizing air which enters the bed at 77 °F and 1 atm are:

Solids: $\rho_s = 62.45$ lb/ft³, $\phi_s = 1.0$, $\varepsilon_{mf} = 0.4$, $\varepsilon_m = 0.5$
Fluid: $\rho_g = 0.0752$ lb/ft³, $\mu = 0.018$ cps.

MA13: For fluidization of the bed, the minimum allowable velocity [ft/s] is nearly

    (A) 0.015   (B) 0.01224   (C) 0.022   (D) 0.025

MA14: The maximum allowable velocity [ft/s] through the bed is

    (A) 0.496   (B) 0.372   (C) 0.248   (D) 0.610

MA15: Fraudie number for the bed based on the mean diameter of the particles is close to

    (A) 0.145   (B) 0.0145   (C) 0.03   (D) 0.29

MA16: Static pressure drop [psi / ft of bed height] is nearly

    (A) 0.26   (B) 0.52   (C) 0.78   (D) 0.34

Properties of fluids:

MA17: The kinematic viscosity of a liquid at 92 °F is 3.71 centistokes. Its density at the same temperature is 54.2 lb/ft³. Its viscosity in engineering units [lb/(ft.h)] is

    (A) 4.2   (B) 5.4   (C) 7.8   (D) 6.7

MA18: The enthalpy [Btu/LB] of superheated steam referred to 32 °F at a pressure of 430 Asia and temperature of 630 °F is nearer to

    (A) 1321.67   (B) 1306.9   (C) 1302.8   (D) 1362.7

MA19: Sato and Riedel give the following relation for estimation of thermal conductivity of a liquid at moderate pressures if critical temperature and boiling point for the liquid are known.

$$k_L = \left[\frac{2.64 \times 10^{-3}}{M^{1/2}} \right] \left[\frac{3 + 20(1-T_r)^{0.67}}{3 + 20(1-T_{br})^{0.67}} \right]$$

    Where k_L = Liquid thermal conductivity, cal/(cm.s.°C)
             M = Molecular weight
             T_r = Reduced temperature
             T_{br} = Reduced boiling point temperature

For benzene, the following data are available: t_c = 289 °C t_b = 80.1 °C.
The thermal conductivity [Btu/h.ft.°F] of benzene at 60 °C is about

    (A) 0.372   (B) 0.0758   (C) 0.1   (D) 0.031

MA20: It is possible to estimate the enthalpy of vaporization at normal boiling point of a liquid by the following equation proposed by Riedel.

$$\Delta H_{vb} = 1.093 R T_c \left[T_{br} \frac{(\ln P_c - 1)}{0.93 - T_{br}} \right]$$

where ΔH_{vb} = heat of vaporization J/gmol

R = gas constant, (atm.cm³)/(gmol.K)
P_c = Critical pressure, atm.
T_c = Critical temperature, K
T_r = reduced temperature
T_{br} = reduced boiling point temperature.

The following data for propyl ether are available from Perry's Handbook;

t_c = 276.2 °C, P_c = 32.9 atm, BP = 101.8 °C, MW= 102.13

The heat of vaporization [Btu/lb] of propyl ether is nearer to

(A) 180   (B) 200   (C) 145   (D) 135

HEAT TRANSFER

Condensation:

MA21 to MA24: 30000 lb/h of a superheated mixture of isobutane/n-butane at 85 psig and 200 °F is condensed in a horizontal condenser. The gas becomes saturated at 126 °F and condenses completely at 125 °F. The cooling water enters the condenser at 65 °F and leaves at 100 °F. The condenser has the following attributes

Shell ID = 23.25", No. of tubes N_t = 352 ¾" OD, ID = 0.62", L_t = 16 ft
Triangular pitch 1", 4 tube passes, Baffle spacing = 12"

Properties of vapor and condensed liquid:

Enthalpy of superheated vapor at 200 °F and 100 psia = 286.7 Btu/lb
Enthalpy of saturated vapor at 126 °F = 231.4 Btu/lb
Enthalpy of the condensed liquid at 125 °F = 119.1 Btu/lb
Density of condensed liquid = 34.3 lb/ft³
Water film coefficient of heat transfer (tube side) = 850 Btu/h.ft². °F

Viscosities of condensed liquid at various temperatures;

Temperature °F	125	120	110	100
Viscosity C_P	0.131	0.134	0.141	0.148

MA21: The heat duty [MM Btu/h] of the condenser is close to
(A) 1.06   (B) 5.03   (C) 3.97   (D) 4.2

MA22: The weighted temperature difference, ΔT [°F] is close to
    (A) 47.9   (B) 52.6   (C) 78.3   (D) 82.5

MA23: If the condensation coefficient of heat transfer based on tube outside area is 150 Btu/h.ft². °F, the condensate film temperature is close to

    (A) 108   (B) 90   (C) 99   (D) 104.8

Hint: Use the relation $t_f = t_{sv} - \frac{3}{4}(t_{sv} - t_w)$ for the calculation of condensate film temperature.

    where t_f = Condensate film temperature, °F
           t_{sv} = Temperature of condensing vapor, °F
           t_w = Tube wall temperature, °F

MA24: Assuming the condensation occurs over a 10 ft of tube length, the condensation coefficient [Btu/h.ft². °F] is close to

    (A) 200   (B) 250   (C) 165   (D) 190

Hint: Use the following relation to calculate the condensation coefficient of heat transfer.

$$\overline{h}\left(\frac{\mu_f^2}{k_f^3 \rho_f^2 g}\right)^{\frac{1}{3}} = 1.5\left(\frac{4G''}{\mu_f}\right)^{-\frac{1}{3}}$$

where $\overline{h}$ = average condensate film coefficient of heat transfer, Btu/h.ft². °F
          μ_f = viscosity of condensate at the film temperature in consistent units,
          ρ_f = denity of condensate at the film temperature, lb/ft³
          g = acceleration due to gravity in consistent units

$$G'' = \frac{W}{L N_t^{\frac{2}{3}}}$$ where G'' = condensate loading lb/h.lin ft

    where W = condensate load, lb/h
             L = Tube length over which condensation occurs
             N_t = Number of tubes

Convection:

MA25: An oil is flowing at a velocity 0f 4.5 ft/s through a 10 ft long, 1" OD, 18 BWG, (ID = 0.902") tube. Steam is condensing on the outside surface of the tube at a temperature of 220 °F. Oil enters the tube at 80 °F and leaves at 100 °F. Properties of oil assumed constant are

    density = 56 lb/lb³, specific heat = 0.48 Btu/lb. °F,
    thermal conductivity = 0.08 Btu/h.ft².°F/ft

The viscosity of oil varies with temperature as follows

t °F	80	90	100	110	120	140	220
μ cP	20	18	16.2	15	13.5	11	3.6

The flow is expected to be streamline. The inside film coefficient of heat transfer [Btu/h.ft². °F] is close to

    (A) 32.2   (B) 30   (C) 34.1   (D) 36.8

MA26 to MA28: 100,000 lb/h of a caustic solution ($\rho = 69.6$ lb.ft³) is to be cooled from 190 to 120 °F. Cooling water at a rate of 154000 lb/h enters the tubes of a 1-4 exchanger at 80 °F and leaves at 120 °F. The dimensions of the exchanger are:

Shell side	tube side
shell ID = 21.25"	$N_t = 172$ 16'0"
Baffle spacing = 6"	Tube OD = 1", 14 BWG, ID = 0.834"
	$1\frac{1}{4}$" triangular pitch
1 pass	4 passes

The properties of the fluids are as follows

Caustic solution:

    Specific heat = 0.88 Btu/lb. °F
    Thermal conductivity = 0.34 Btu/h.ft². °F/ft
    Viscosities at different temperatures:

t °F	80	90	100	120	140	160	180	200	210	220
μ cP			1.4						0.43	

Water:

t °F	80	90	100	120	140	160	180
μ lb/h.ft	2.08	1.85	1.66	1.36	1.14	0.970	0.840
k Btu/h.ft². °F/ft	0.351	0.357	0.363	0.372	0.379	0.385	0.390

MA26: The tube heat transfer coefficient [Btu/h.ft². °F] based on the inside surface is most near to

    (A) 900   (B) 1050   (C) 1500   (D) 850

Hint: Use Dittus - Boelter equation to calculate the heat transfer coefficient.

MA27: The mass velocity on the shell side [lb/h.ft²] is most nearly

    (A) 564700   (B) 250000   (C) 350000   (D) 850000

MA28: The shell side heat transfer coefficient [Btu/h.ft². °F] is most near to

(A) 900   (B) 1000   (C) 770   (D) 850

Fouling:

MA29: For a countercurrent heat exchanger, inside and outside heat transfer coefficients h_i and h_o have been computed as 700 and 250 Btu/h.ft².°F respectively. By experience, it is known that a dirt resistance, $R_{di} = 0.001$ will deposit annually on the inside of the tube while $R_{do} = 0.0015$ will deposit on the outside of the tube. The tube wall thickness is very small. The metal wall resistance can be neglected. The overall heat transfer coefficient [Btu/h.ft².°F] on the basis of which the heat transfer surface should be calculated so that the exchanger need be cleaned only once a year is nearer to

(A) 100   (B) 124   (C) 172   (D) 150

MA30 to MA32: Overall heat transfer coefficients based on outside area were determined in series of experiments for the condensation of steam on the outside of a dirty and a clean tube with water flowing through the tubes at various velocities. Wilson plots of the two sets of data are plotted as $1/U_0$ vs $1/u^{0.8}$ in figure below. Assume constant steam film coefficient on the outside of the tubes. Tubes are admiralty-metal, 1"OD and 0.902" ID. Thermal conductivity of tube metal = 63 Btu/h.ft².°F/ft.

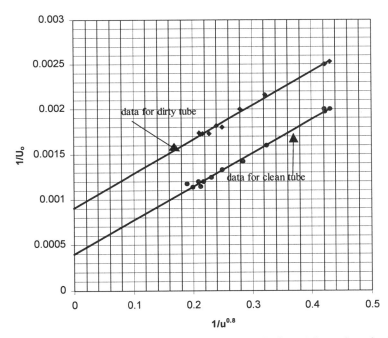

MA30: The water side heat transfer coefficient [Btu/h.ft².°F] based on inside area of the tubes at a velocity of 1 ft/s is close to

(A) 246   (B) 263   (C) 252   (D) 292

MA31: The condensation coefficient [Btu/h.ft². °F] for steam based on outside area of tube is nearer to

    (A) 3012   (B) 2750   (C) 2500   (D) 2000

MA32: The scale deposit coefficient h_d [Btu/h.ft².°F] based on inside area of the tube is most nearly

    (A) 2700   (B) 2500   (C) 2250   (D) 2130

Insulation:

MA33 to MA36: A standard 2" pipe carrying superheated steam is insulated with the following:

    First layer : 1.25" thk diatomaceous earth + asbestos, k = 0.058 Btu/h.ft².°F
    Second layer: 2.5" thk laminated asbestos felt,   k = 0.042 Btu/h.ft².°F

    Other data: Temperature of surroundings = 80 °F
                  Average temperature of steam pipe = 900 °F
    Temperature of outer surface of second layer = 120 °F
    Thermal conductivity of steel = 26 Btu/h.ft².°F
    Mean diameter of pipe = 2.222 in
    Inside diameter of pipe = 2.067 in
    Outside diameter of pipe = 2.375"
    OD of first layer of insulation = 4.875"
    OD of second layer of insulation = 9.87"
    Log mean diameter of first layer of insulation = 3.48"
    Log mean diameter of second layer of insulation = 7.07"

MA33: Total resistance [h.ft².°F/Btu] of the two layers of insulation is close to

    (A) 4.651   (B) 2.68   (C) 1.97   (D) 2.33

MA34: Heat loss per foot length of pipe [Btu/h.lin.ft of pipe] is nearer to

    (A) 162   (B) 181   (C) 168   (D) 152

MA35: Interface temperature [°F] between the two insulation layers is nearer to

    (A) 569   (B) 352   (C) 446   (D) 331

MA36: The radiation heat transfer coefficient [Btu/h.ft².°F] is close to

    (A) 0.87   (B) 0.6   (C) 1.0   (D) 1.2

PLANT DESIGN

Economics:

MA37 to MA38: A project has the following cost data

> Initial investment = $ 120,000
> Working capital = $ 15,000
> Salvage value = $ 12,000
> Project useful life = 5 years

Annual cash flows to the project after taxes (based on total income less all costs except depreciation) are as follows

> First year $ 35,000
> Second year $ 37,000
> Third year $ 43,000
> Fourth year $ 48,000
> Fifth year $ 52,000

Additional data: (Discount factors)

Table PA1
Factor = e^{-rn}

Factors for present worth calculation based on discrete interest compounding and instantaneous cash flows

Interest rate %	0.1	0.15	0.2	0.25
0-1 year	0.9516	0.9286	0.9063	0.8848
1-2 year	0.8611	0.7993	0.7421	0.6891
2-3 year	0.7791	0.6879	0.6075	0.5367
3-4 year	0.7000	0.5921	0.4974	0.4179
4-5 year	0.6400	0.5096	0.4072	0.3255

$$Factor = (\frac{e^r - 1}{r}) e^{-rn}$$

Factors for present worth calculation based on continuous interest compounding and continuous cash flows

Interest rate %	0.1	0.15	0.2	0.25
n = 1	0.9048	0.8607	0.8187	0.7788
2	0.8187	0.7408	0.6703	0.6065
3	0.7408	0.6376	0.5488	0.4724
4	0.6703	0.5488	0.4493	0.3679
5	0.6065	0.4724	0.3679	0.2865

MA37: The percent rate of return based on discounted cash flow with discrete interest compounding and the cash flow compounded on the basis of end-of-year income is nearer to
   (A) 19   (B) 20   (C) 18   (D) 22.5

MA38: The percent rate of return with continuous interest compounding and continuous cash flow using present worth of all cash flows and the factors given in table PA1 is nearer to

(A) 22.6 (B) 19.7 (C) 18.8 (D) 21.7

MA39: The estimated annual fixed costs and the costs incurred due to heat loss are plotted in the following figure.

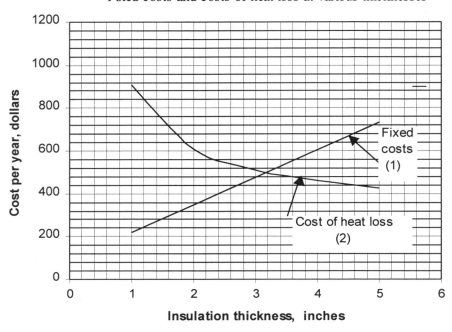

These data indicate an optimum insulation thickness [inches] of

(A) 3.25 (B) 5.2 (C) 2.25 (D) 4

MA40: A reactant A is converted to a product R in a liquid phase reaction. A also decomposes to another undesirable product S. The reactions are irreversible and take place as follows

$$A \xrightarrow{k_1} R \qquad k_1 = 4C_A$$

$$A \xrightarrow{k_2} S \qquad k_2 = 1C_A$$

It is required to process a feed containing only A in a reactor of volume 60 liters. Cost data are :

$C_{A0} = 1$ mol/liter
Feed A, cost = \$1.5/mol
Product R, cost = \$5/mol

Product waste, S has no value. Unreacted A is separated and recycled
Operating costs = $30 + $1.5/mol A per hour

For a mixed reactor with first order kinetics, the volume of the mixed reactor is given by

$$V = \frac{F_{Ao} X_A}{k C_{Ao}(1 - X_A)}$$

where F_{Ao} = molar feed per hour of reactant

X_A = conversion of A

The % conversion to maximize profits is near to

        (A) 95   (B) 77.5   (C) 91.4   (D) 68.5

STOP. CHECK YOUR WORK. END OF PM EXAM 1..

AFTERNOON SAMPLE EXAMINATION

Instructions for afternoon Session

1. You have four hours to work on the afternoon session. Do not write in this handbook.

2. Answer all forty questions for a total of forty answers. There is no penalty for guessing.

3. Work rapidly and use your time effectively. If you do not know the correct answer, skip it and return to it later.

4. Some problems are presented in both metric and English units. Solve either problem.

5. Mark your answer sheet carefully. Fill in the answer space completely. No marks on the workbook will be evaluated. Multiple answers receive no credit. If you make a mistake, erase completely.

Work all 40 problems in four hours.

P.E. Chemical Engineering Exam
Afternoon Session

Ⓐ Ⓑ Ⓒ Fill in the circle that matches your exam booklet.

AA1 Ⓐ Ⓑ Ⓒ Ⓓ	AA11 Ⓐ Ⓑ Ⓒ Ⓓ	AA21 Ⓐ Ⓑ Ⓒ Ⓓ	AA31 Ⓐ Ⓑ Ⓒ Ⓓ
AA2 Ⓐ Ⓑ Ⓒ Ⓓ	AA12 Ⓐ Ⓑ Ⓒ Ⓓ	AA22 Ⓐ Ⓑ Ⓒ Ⓓ	AA32 Ⓐ Ⓑ Ⓒ Ⓓ
AA3 Ⓐ Ⓑ Ⓒ Ⓓ	AA13 Ⓐ Ⓑ Ⓒ Ⓓ	AA23 Ⓐ Ⓑ Ⓒ Ⓓ	AA33 Ⓐ Ⓑ Ⓒ Ⓓ
AA4 Ⓐ Ⓑ Ⓒ Ⓓ	AA14 Ⓐ Ⓑ Ⓒ Ⓓ	AA24 Ⓐ Ⓑ Ⓒ Ⓓ	AA34 Ⓐ Ⓑ Ⓒ Ⓓ
AA5 Ⓐ Ⓑ Ⓒ Ⓓ	AA15 Ⓐ Ⓑ Ⓒ Ⓓ	AA25 Ⓐ Ⓑ Ⓒ Ⓓ	AA35 Ⓐ Ⓑ Ⓒ Ⓓ
AA6 Ⓐ Ⓑ Ⓒ Ⓓ	AA16 Ⓐ Ⓑ Ⓒ Ⓓ	AA26 Ⓐ Ⓑ Ⓒ Ⓓ	AA36 Ⓐ Ⓑ Ⓒ Ⓓ
AA7 Ⓐ Ⓑ Ⓒ Ⓓ	AA17 Ⓐ Ⓑ Ⓒ Ⓓ	AA27 Ⓐ Ⓑ Ⓒ Ⓓ	AA37 Ⓐ Ⓑ Ⓒ Ⓓ
AA8 Ⓐ Ⓑ Ⓒ Ⓓ	AA18 Ⓐ Ⓑ Ⓒ Ⓓ	AA28 Ⓐ Ⓑ Ⓒ Ⓓ	AA38 Ⓐ Ⓑ Ⓒ Ⓓ
AA9 Ⓐ Ⓑ Ⓒ Ⓓ	AA19 Ⓐ Ⓑ Ⓒ Ⓓ	AA29 Ⓐ Ⓑ Ⓒ Ⓓ	AA39 Ⓐ Ⓑ Ⓒ Ⓓ
AA10 Ⓐ Ⓑ Ⓒ Ⓓ	AA20 Ⓐ Ⓑ Ⓒ Ⓓ	AA30 Ⓐ Ⓑ Ⓒ Ⓓ	AA40 Ⓐ Ⓑ Ⓒ Ⓓ

PM SAMPLE EXAM 2
FLUIDS
Control of flow systems:

AA1 to AA4: The table below is an extract from the files of a process engineer's calculations for the discharge side of a pump.

		Max flow Branch		Normal flow Branch		Minimum flow Branch	
		#1	#2	#1	#2	#1	#2
Flow	m³/h	12.5	17.4	10.2	14.5	4.2	5.8
Delivery pressure	kg/cm² a	2.0	3.5	2.0	3.5	2.0	3.5
Static head	kg/cm² a	3.72	0.00	3.72	0.00	3.72	0.00
Line losses	kg/cm²	0.81		0.56	0.42	0.09	0.07
equipment losses	kg/cm²	0.61		0.42	1.68	0.05	0.27
Control valve ΔP	kg/cm²	0.75		1.19		2.03	
Discharge Pressure	kg/cm² a	7.89	7.89	7.89	7.89	7.89	7.89
Suction pressure	kg/cm² a	2.17	2.17	2.18	2.18	2.19	2.19
Pump differential	kg/cm²	5.72					
Differential head,	m	72.6					

AA1: The required rangeability of the control valve to be installed in flow Branch #1 is close to:

(A) 10:1 (B) 5:1 (C) 12:1 (D) 8:1

AA2: The pressure drop [kg/cm²] to be allowed for the control valve in Branch #2 at the normal flow is close to:

(A) 1.03 (B) 0.7 (C) 2.29 (D) 1.17

AA3: The control valve pressure drop [kg/cm²] at the maximum flow in Branch #2 will be close to:

(A) 1.03 (B) 0.7 (C) 0.8 (D) 1.36

AA4: A manufacturer's catalog lists flow coefficients for his standard line of valves as follows:

Valve size, inches	1	1½	2	3	4	6
Flow coefficient, C_V	9	21	36	75	124	270

The most suitable valve size [in] for the control valve in Branch #1 is:

(A) 2 (B) 1½ (C) 1 (D) 3

Economics:

AA5 to AA8: The installed cost of an equipment is $100,000. Its salvage value is estimated at 10 % of its initial cost. Its service life is expected to be 10 years.

AA5: If straight line depreciation is allowed, the book asset value of the equipment after 5 years would be

(A) 50,000   (B) 45,000   (C) 64,000   (D) 55,000

AA6: If declining balance depreciation is used, the book value [$] after 5 years will be

(A) 31617   (B) 68383   (C) 60454   (D) 37,262

AA7: If the method of double declining balance is used, the book value [$] after 5 years will be

(A) 51,342   (B) 45,623   (C) 32768   (D) 62,256

AA8: If the depreciation is calculated by the sum-of-digits method, the book value [$] after 5 years will be nearer to

(A) 65455   (B) 45,365   (C) 34545   (D) 24,545

Sensors:

AA9: Water is flowing in a 3" schedule 40 pipe. at a rate of 34.1 m^3/h. Sp gr of water = 1 and viscosity = 1 cP. A 1¾" dia orifice is mounted in the pipe to measure the water flow. A mercury manometer connected across the orifice will show a differential height [cmHg] most nearly

(A) 40.5   (B) 38.1   (C) 36.2   (D) 30.6

AA10: A pneumatic pressure controller is calibrated to a pressure range of 0 to 60 kg/cm^{2}g. The signal output is 70 kPa. The indicated pressure [kg/cm^{2}g] is close to

(A) 57.5   (B) 45.6   (C) 37.5   (D) 42.5

AA11: A differential pressure transmitter is calibrated 0 to 220 cms of water column and transmits 4 to 20 mA dc signal. The transmitter is attached to an orifice sized to create 220 cm of differential of water column at a flow rate of 15 m^3/h. The flow rate [m^3/h] when the transmitter signal is 15 mA dc is nearly

(A) 10   (B) 12.5   (C) 15   (D) 17.0

AA12: A thermocouple has a time constant of 2 seconds. If the process temperature changes suddenly from 900 to 1000 °C, the temperature read out of the indicator attached to the thermocouple after 6 seconds elapsed time will be close to

(A) 935   (B) 995   (C) 970   (D) 1000

Pumps and turbines:

AA13 to AA16: The sketch below shows a pumping system to transfer toluene-xylene mixture from stabilizer to the toluene column.

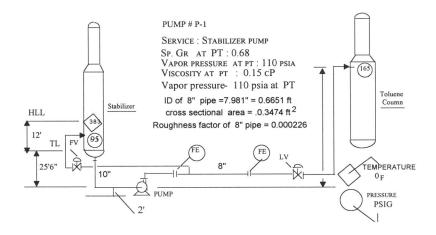

Some data are given in the sketch. Some other calculations are given in the following table:

	Rated	Normal
Flow rate gpm	1452	1320
Suction side:		
Origin pressure psia	110	110
Suction static head psi (based on bottom tangent line of stabilizer)	--	--
Line loss psi	1.5	1.2
Pump suction pressure psia		
Vapor pressure at PT psia		
NPSH available psi		
NPSH available ft		
NPSH Design (To be specified) ft		
Discharge side:		
Delivery pressure psia	180.0	180.0
Discharge static head psi	55.4	55.4
Line losses psi	7.3	6.0
Control valve pressure drop psi	10.4	15.0
Misc. equipment losses psi	12.1	10.0
Orifice pressure drop (2 Nos.) psi	4.8	4.0
Dynamic losses psi	2.4	2.0
Pump discharge pressure psia		
- Pump suction pressure psia		

AA13: Suction static head [psi] is most nearly:

(A)   7.5   (B)   6.6   (C)   10.8   (D)   11.0

AA14: Based on tangent line of stabilizer column, NPSH [ft] available at the rated flow is:

(A) 20.4 (B) 18.4 (C) 25 (D) 16.0

AA15: Total dynamic head [ft] is most nearly

(A) 362 (B) 550 (C) 480 (D) 531

AA16: If the efficiency of the pump is 65 %, the brake horse power of the pump is most nearly:

(A) 204 (B) 133 (C) 145 (D) 220

HEAT TRANSFER

Resistance:

AA17 to AA18: The temperature profile across a composite wall is as shown in the sketch below. The thermal coductivities are also given in the sketch.

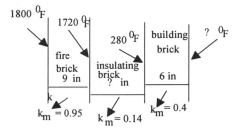

k_m in Btu/(h.ft^2. ^{0}F/ft)

AA17: If the furnace has 450 ft^2 of wall surface, the heat loss [Btu/day] by conduction through the wall is most nearly

(A) 3.2x10^6 (B) 0.68x10^6 (C) 2.3x10^5 (D) 1.64x10^6

AA18: The thickness [inches] of the insulating brick is most nearly

(A) 16 (B) 12 C) 14 (D) 10

AA19 to AA20: The following data were obtained from a heat exchanger:

  Tube side film coefficient (for water) = 1840 Btu/h.ft^2.^{o}F
  Shell side film coefficient (for steam) = 2010 Btu/h.ft^2.^{o}F
  Fouling factor outside tube   = 0.0005 h.ft^2.^{o}F/Btu
  Fouling factor inside tube   = 0.001 h.ft^2.^{o}F/Btu
  Tube   ID = 0.902"   OD = 1" Thickness = 0.049" k_m = 63 Btu/(h.ft^2.^{o}F/ft)

AA19: Based on the outside surface, the overall heat transfer coefficient U$_o$ [Btu/h.ft^2.^{o}F] will be:

(A) 361 (B) 860 (C) 600 (D) 454

AA20: If the tubes are cleaned on both sides to remove the scale, the clean overall coefficient, U_o [Btu/h.ft^2.°F] will be

(A) 680   (B) 860   (C) 730   (D) 800

Conduction:

AA21 to AA24: A standard 2" steel pipe is carrying steam at 900 °F. The pipe is lagged with 1½" magnesia covering outside of which there is a 2.5" of insulating cork. Outside surface temperature is 90 °F. Thermal conductivities in units of Btu/h.ft^2. °F/ft are:

steel k = 26   magnesia k = 0.054 and cork k = 0.03

Surrounding temperature is 86 °F. ID of pipe = 2.067" OD of pipe = 2.375"

AA21: Resistance of pipe wall [(h.ft^2. °F/Btu] is nearly

(A) 0.00085   (B) 0.0009   (C) 0.00075   (D) 0.00080

AA22: Heat loss [Btu/h.ft of pipe] per foot of pipe is nearly

(A) 115   (B) 107   (C) 110   (D) 104

AA23: Total resistance to heat transfer is nearly

(A) 4.83   (B) 7.313   (C) 6.22   (D) 5.17

AA24: The temperature [°F] at the interface of magnesia and cork insulations is

(A) 408   (B) 440   (C) 492   (D) 416

Energy conservation:

AA25 to AA26: A plant is presently cooling stripper bottoms by cooling water before discharging to sewer. It is contemplated to use the stripper bottoms to heat the feed and save on steam consumption and cooling water. The present and new schemes are shown in the figure on the next page.

The other data are as follows:

         Operating time = 8000 h/yr
         Installed cost of present exchanger = $ 10000,
         Surface area A = 50 ft^2, U = 100 btu/h.ft^2.°F
         Steam cost $5/1000 lb
         Specific heat of feed or stripper bottoms = 1 Btu/lb·°F (assume)

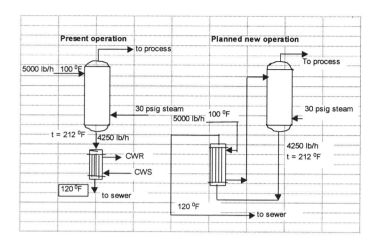

AA25: Yearly saving [$] in steam alone is near to

(A) 16840 (B) 20000 (C) 18320 (D) 12800

AA26: Neglecting piping costs and considering only investment in the new exchanger, the payout period [Months] is near to

(A) 13.7 (B) 15 (C) 12 (D) 10

AA27: A chemical manufacturing plant produces an organic liquid waste as byproduct which is environmentally hazardous. The liquid byproduct has an average lower heating value of 6000 Btu/lb and is completely oxidized at 1800 °F. It is proposed to utilize this value to produce steam at 300 psig and 422 °F and at the same time to eliminate the environmental hazard. It is estimated that a total of 21000 cfm of flue gas (at 1800 °F) will be produced and quenched to 200 °F before sending to a scrubber. Flue gas weight rate will be 178500 lb/h. and its specific heat can be taken as 0.3 Btu/lb.°F. The waste heat boiler was designed to give a flue gas temperature of 500 °F. Assume heat losses of 1 %. Steam cost can be taken as $2.5 lb/1000 lb. Operating time per year = 6000 hours. The net annual savings [$] in steam costs because of this change will be near to

(A) 1018000 (B) 800000 (C) 500000 (D) 900000

AA28: A 6" horizontal pipe carries saturated steam at 175 psig steam. The ambient temperature is 70 °F. The pipe is insulated with 1" insulation which has an emmissivity of 0.67 and thermal conductivity of 0.05 Btu/h.ft².°F/ft. On monitoring the performance of the insulation, it was observed that the heat loss is 2455 MM Btu/yr amounting to $6547/yr fuel expense at $2/10⁶ Btu with the boiler system operating at 75 % efficiency. The length of the pipe line is 1000 ft. Operating time per year is 8760 hours. If the insulation thickness is increased by 2", the acceptable cost of additional insulation [$/linear foot] for a payback period of 2 years is near to

(A) 5.2 (B) 4.6 (C) 6.79 (D) 8

[Hint: The convection heat transfer coefficient h_c for horizontal pipe can be calculated by the simple equation

$$h_c = 0.5\left(\frac{\Delta t}{d_o}\right)^{0.25} \text{ where } d_o \text{ is in inches.}$$

And radiation coefficient can be calculated by the equation

$$h_r = \frac{0.174\epsilon\left[\left(\frac{T_s}{100}\right)^4 - \left(\frac{T_a}{100}\right)^4\right]}{t_s - t_a}$$

Where T is in °R and h_c and h_r are in Btu/h.ft².°F

Radiation:

AA29: A 2" IPS steel pipe carrying steam at 300 °F passes through a duct of galvanized sheet iron which is maintained at 70 °F. The cross section of the duct is 1 ft x 1 ft. The duct is insulated on the outside. Emissivity of steel, $\epsilon = 0.8$. Emissivity of galvanized sheet iron, $\epsilon = 0.28$. The heat loss [Btu/h.linear ft of pipe] is most nearly

   (A) 150 (B) 130 (C) 166 (D) 185

AA30: A bare 2" steel pipe carrying steam at 325 °F passes through a room which is at 70 °F. The % decrease in radiation that occurs if the bare pipe is coated with aluminum paint is nearer **to**

   (A) 50.2 (B) 45.3 (C) 55.7 (D) 60.1

  Data: Emissivity of bare pipe = 0.8
      Emissivity of painted pipe = 0.35

AA31: An oxidized nickel tube of 3" OD and at a temperature of 850 °F is enclosed in a gray chamber 10" square inside. The chamber is lined with glazed silica brick and is maintained at 1850 °F. The emissivity data are as follows

     Glazed silica $\epsilon = 0.78$
     Nickel at 850 °F = 0.4402
     Nickel at 1850 °F = 0.59

The radiative heat transfer [Btu/h.ft²] between the tube and the chamber is most nearly

   (A) 14750 (B) 24883 (C) 20350 (D) 18230

Hint: The overall interchange factor for gray surfaces is given by the relation

$$\frac{1}{A_1 \mathcal{F}_{1\to 2}} = \frac{1}{A_1}\left(\frac{1}{\epsilon_1} - 1\right) + \frac{1}{A_2}\left(\frac{1}{\epsilon_2} - 1\right) + \frac{1}{A_1 \bar{F}_{1\to 2}}$$

where $\mathcal{F}_{1\to 2}$ = overall interchange factor for gray surface
   $\bar{F}_1$ = view factor between surfaces A_1 and A_2 with allowance for refractory surface.
   A_1, A_2 = areas of non-black source-sink surfaces
   ϵ_1, ϵ_2 = emissivities of the source- sink surfaces

AA32: A plot of $\lambda -$ intensity of monocromatic emission for a body at a temperature of 2700 °F in units of Btu/(h.ft².)(micron) versus λ - the wave length in microns is shown in the following graph.

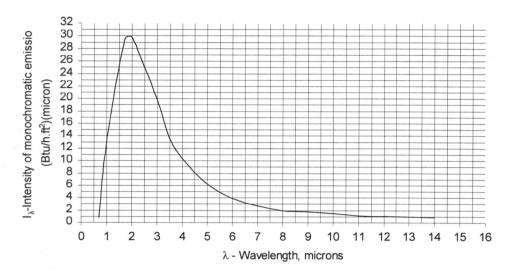

Figure for problem AA32: A plot of monocromatic radiation vs wavelength for a hot body at 2700 °F

The total emissive power of the hot body is closer to

(A) 50.2   (B) 45.0   (C) 55.0   (D) 60.1

MASS TRANSFER

Flooding and pressure drop:

AA33 to AA36: A sieve tray column is specified to handle an aromatic separation. The operating conditions are as follows

Vapor flow rate:

$Q_V = 60$ ft³/s, $\rho_v = 0.256$ lb/ft³,   Flow = 55300 lb/h

Liquid flow rate:

$Q_L = 120.2$ gpm., $\rho_L = 47.8$ lb/ft³, Flow = 46100 lb/h, $\sigma = 18$ dynes/cm

The tower and tray characteristics specified are as follows:

Tower diameter, $D_T = 5$ ft, Number of trays = 40
Hole diameter = ¼", Weir height = 2", Plate thickness = 0.0825", Spacing = 24"
Weir length is to be $0.7267 D_T$ where D_T is the tower diameter.
Ratio of hole area to total tray cross section = $A_h/A = 0.1$

AA33: The head loss due to liquid depth and crest over the weir is near to

    (A) 1.77 (B) 1.4 (C) 1.0 (D) 1.2

AA34: Dry tray pressure drop [in] is near to

    (A) 2 (B) 1.32 (C) 1.73 (D) 2.3

AA35: Pressure drop [in] due to surface tension effect is near to

    (A) 0.2 (B) 0.14 (C) 0.1 (D) 0.06

AA36: The downcomer head loss is most near to

    (A) 0.2 (B) 0.12 (C) 0.18 (D) 0.24

THERMODYNAMICS

Economics:

AA37 to AA38: Mr A takes a 30 year mortgage loan of $200,000 from a savings and loan bank at 6½ % for the purchase of house. The interest is compounded monthly. The first payment is due after one month from the signing date of the contract.

AA37: The effective annual interest rate [%] will be nearer to

    (A) 6.0 (B) 6.5 (C) 6.2 (D) 7.0

AA38: The amount of monthly payment [$] will be nearer to

    (A) 1264.19 (B) 1120.58 (C) 1359.22 (D) 1087.73

AA39: The original cost of a property is 40,000 dollars. and it is depreciated by using sinking-fund factor method. The interest rate assumed is 6½%. If the book value of the property after 10 years is the same as if it has been depreciated at $3350/yr by straight line method, the annual depreciation charge [$] per year is nearer to

    (A) 3335.10 (B) 3034.23 (C) 2871.56 (D) 2482.51

AA40: The capitalized cost of an equipment with zero salvage value and 10 years useful life was found to be $100,000. This cost was based on the original cost plus the present value of an indefinite number of renewals. A 10% interest rate was assumed in the calculations. The original cost of the equipment would be

    (A) 65219.12 (B) 74150.34 (C) 61445.67 (D) 52336.4

STOP. CHECK YOUR WORK. END OF PM EXAM 2..

AFTERNOON SAMPLE EXAMINATION

Instructions for afternoon Session

1. You have four hours to work on the afternoon session. Do not write in this handbook.

2. Answer all forty questions for a total of forty answers. There is no penalty for guessing.

3. Work rapidly and use your time effectively. If you do not know the correct answer, skip it and return to it later.

4. Some problems are presented in both metric and English units. Solve either problem.

5. Mark your answer sheet carefully. Fill in the answer space completely. No marks on the workbook will be evaluated. Multiple answers receive no credit. If you make a mistake, erase completely.

Work all 40 problems in four hours.

P.E. Chemical Engineering Exam
Afternoon Session

Ⓐ Ⓑ Ⓒ Fill in the circle that matches your exam booklet.

MB1 Ⓐ Ⓑ Ⓒ Ⓓ	MB11 Ⓐ Ⓑ Ⓒ Ⓓ	MB21 Ⓐ Ⓑ Ⓒ Ⓓ	MB31 Ⓐ Ⓑ Ⓒ Ⓓ
MB2 Ⓐ Ⓑ Ⓒ Ⓓ	MB12 Ⓐ Ⓑ Ⓒ Ⓓ	MB22 Ⓐ Ⓑ Ⓒ Ⓓ	MB32 Ⓐ Ⓑ Ⓒ Ⓓ
MB3 Ⓐ Ⓑ Ⓒ Ⓓ	MB13 Ⓐ Ⓑ Ⓒ Ⓓ	MB23 Ⓐ Ⓑ Ⓒ Ⓓ	MB33 Ⓐ Ⓑ Ⓒ Ⓓ
MB4 Ⓐ Ⓑ Ⓒ Ⓓ	MB14 Ⓐ Ⓑ Ⓒ Ⓓ	MB24 Ⓐ Ⓑ Ⓒ Ⓓ	MB34 Ⓐ Ⓑ Ⓒ Ⓓ
MB5 Ⓐ Ⓑ Ⓒ Ⓓ	MB15 Ⓐ Ⓑ Ⓒ Ⓓ	MB25 Ⓐ Ⓑ Ⓒ Ⓓ	MB35 Ⓐ Ⓑ Ⓒ Ⓓ
MB6 Ⓐ Ⓑ Ⓒ Ⓓ	MB16 Ⓐ Ⓑ Ⓒ Ⓓ	MB26 Ⓐ Ⓑ Ⓒ Ⓓ	MB36 Ⓐ Ⓑ Ⓒ Ⓓ
MB7 Ⓐ Ⓑ Ⓒ Ⓓ	MB17 Ⓐ Ⓑ Ⓒ Ⓓ	MB27 Ⓐ Ⓑ Ⓒ Ⓓ	MB37 Ⓐ Ⓑ Ⓒ Ⓓ
MB8 Ⓐ Ⓑ Ⓒ Ⓓ	MB18 Ⓐ Ⓑ Ⓒ Ⓓ	MB28 Ⓐ Ⓑ Ⓒ Ⓓ	MB38 Ⓐ Ⓑ Ⓒ Ⓓ
MB9 Ⓐ Ⓑ Ⓒ Ⓓ	MB19 Ⓐ Ⓑ Ⓒ Ⓓ	MB29 Ⓐ Ⓑ Ⓒ Ⓓ	MB39 Ⓐ Ⓑ Ⓒ Ⓓ
MB10 Ⓐ Ⓑ Ⓒ Ⓓ	MB20 Ⓐ Ⓑ Ⓒ Ⓓ	MB30 Ⓐ Ⓑ Ⓒ Ⓓ	MB40 Ⓐ Ⓑ Ⓒ Ⓓ

PM SAMPLE EXAM 3
FLUIDS
Economics:

MB1 to MB4: A proposed manufacturing plant requires an initial fixed capital investment of $1.1 Million, and $120,000 as working capital. Estimated annual income will be $960,000. Annual expenses including depreciation are $620,000. A minimum of 15% return before taxes is required for the investment to be worthwhile. Income tax is 40% of gross profits before taxes.

MB1: Annual percent return before income taxes is nearer to
  (A) 30.1   (B) 27.9   (C) 25.3   (D) 28.6

MB2: Annual percent return on the initial investment after income taxes is near to
  (A) 18.2   (B) 16.4   (C) 13.2   (D) 14.49

MB3: Annual % return on the initial investment before income taxes based on capital recovery with minimum profit is nearer to
  (A) 15.4   (B) 13.8   (C) 14.6   (D) 12.9

MB4: The annual percent return on the average investment before income taxes assuming straight line depreciation and zero salvage value is nearer to.
  (A) 48   (B) 50.75   (C) 45.4   (D) 52.8

KINETICS
Biochemical reactors:

MB5 to MB6: Data on the hydrolysis of sucrose at room temperature are plotted in the following figure with $-1/r_a$ as ordinate and $1/C_A$ as abscissa.

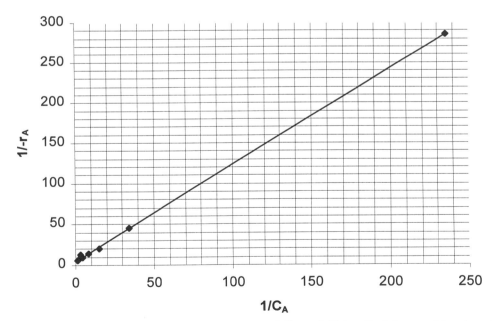

The plot indicates that the hydrolysis of sucrose follows Michaelis-Menton kinetics given by the equation

$$-r_A = \frac{k_3 C_A C_{E0}}{C_A + K_M}$$

- $-r_A$ = rate of hydrolysis of sucrose to products
- k_3 = reaction rate constant, h^{-1}
- C_A = concentration of sucrose, millimol/liter at any time during hydrolysis
- M = Michaelis constant

MB5: The value of the constant k_3 [h^{-1}] is nearer to
    (A) 20.95   (B) 15.3   (C) 18.7   (D) 23.12

MB6: The value of the Michaelis-Menton constant is close to
    (A) 0.4   (B) 0.5   (C) 0.25   (D) 0.35

MB7 to MB8: Activity of an enzyme having K_m of 1×10^{-3} M was tested using an initial substrate concentration of 3×10^{-5} M. After 2 min., 5% of the substrate was converted.

MB7: Maximum rate of substrate conversion [M/(min.unit vol.)] will be close to
    (A) 2.0×10^{-5} (B) 1.7×10^{-5} (C) 2.64×10^{-5} (D) 3.1×10^{-5}

MB8: The substrate conversion [%] after 30 min. will be nearer to
    (A) 54   (B) 50   (C) 67   (D) 46

Equilibrium: chemical/phase:

MB9 to MB12: The following reaction takes place at 704 °C and 2 atm.

$$C_4H_{10}(g) = 2C_2H_4(g) + H_2(g)$$

The data on ΔG^o, ΔH^o at 298 K and C_p values are given in the following table.

Component	ΔG^0_{298} kcal/gmol	ΔH^o_{298} kcal/gmol	C_P kcal/gmol.K
C_4H_{10}	-3.75	-29.81	$C_p = 0.01178 + 4.268 \times 10^{-5} T$
C_2H_4	16.28	12.5	$C_p = 0.0028 + 3.000 \times 10^{-5} T$
H_2	0	0	$C_p = 0.0069 + 0.4 \times 10^{-5} T$

[Hints: When component C_ps are expressed by equations of the type

$$C_p^o = \alpha + \beta T + \gamma T^2 + \cdots$$

One can evaluate net C_p^o for products and reactants as given below.

$$\Delta C_p^o = \Delta\alpha + \Delta\beta T + \Delta\gamma T^2 + \cdots$$

where $\Delta\alpha = (\Sigma n\alpha)_{products} - (\Sigma n\alpha) reactants$ with similar equations for $\Delta\beta$ and $\Delta\gamma$ etc

Then various relations have been developed as follows

Heat of reaction at any temperature,

$$\Delta H_T^o = I_H + \Delta a T + \Delta \beta (\tfrac{1}{2}) T^2 + \Delta \gamma (\tfrac{1}{3}) T^3 + \cdots$$

Equilibrium constant, K:

$$\ln K = -\frac{I_H}{RT} + \frac{\Delta a}{R} \ln T + \frac{\Delta \beta}{R}(\tfrac{1}{2})T + \frac{\Delta \gamma}{R}(\tfrac{1}{6})T^2 + \cdots + I$$

where I is another constant which can be evaluated if one value of K is known. Also, the following relation gives free energy of a reaction at any temperature T

$$\Delta G_T^o = I_H + I_G T - \Delta a T - \tfrac{1}{2}\Delta \beta T^2 - \tfrac{1}{6}\Delta \gamma T^3 + \cdots$$

where $I_{PG} = -(IR)$

MB9: The free energy [kcal/gmol] of the reaction at 25 °C is near to
 (A) 36.31 (B) -36.31 (C) 20.03 (D) -20.3

MB10: Heat [kcal/gmol] of this reaction at a temperature of 500 °C is near to
 (A) -60.57 (B) 60.57 (C) -29.93 (D) +29.93

MB11: The equilibrium constant K for the reaction at temperature of 704 °C is closer to
 (A) 6.5 (B) 133 (C) 303.7 (D) 62.3

MB12: The equilibrium conversion [%] of n-butane at temperature of 704 °C is nearer to
 (A) 99.56 (B) 65.53 (C) 80.21 (D) 91.17

Material selection for reactors:

MB13: In industrial production of high grade phosphoric acid, liquid phosphorus is sprayed in an oxidation tower and oxidized to phosphorus pentoxide (P_2O_5). The oxide is cooled and hydrated to phosphoric acid using dilute phosphoric acid as absorption medium. A suitable material of construction for the hydrator producing 75 % phosphoric acid is
 (A) Carbon steel
 (B) SS 317L
 (C) SS 430
 (D) Aluminum.

MB14: Alkylation of dodecene with benzene is carried out in presence of $AlCl_3$ as catalyst. The material of construction for this reactor can be

 (A) SS 304
 (B) SS 316
 (C) Alloy 20
 (D) Glass-lined steel

MB15: A chemical company produces chlorinated and fluorinated chemicals. Reactor pressure vessels are constructed of alloy C-276 (Ni-57 %, Cr = 16 % MO = 16 %, Fe = 5 %). Because of excessive corrosion, they have to be replaced every 12 to 14 months. The corrosion rate of C-276 in 10 % H_2SO_4 + 1 % HCl solution at 90 °C is 0.041 in. per year while that of alloy 59 is 0.003 in per year. The process involves chemicals such as hydrocarbons, ammonium fluoride, sulfuric acid and others. A suitable new material of construction for the reactor in order to have prolonged life for the same when compared with C-276 would be

(A) Glassed lined steel
(B) Tantalum clad vessel
(C) Alloy 59 (Ni = 59 %, Cr = 23 %, Mo = 16 % , Fe = 1%)
(D) SS 317L

MB16: In steam reforming of hydrocarbons, a mixture of steam and hydrocarbon is catalytically reacted at about 1400-1800 °F in a tubular reactor, followed by a shift reaction in which CO and steam react to produce CO_2 and hydrogen. The catalyst is nickel. The process involves removal of sulfur from the hydrocarbon to about 5 ppm level, mixing with superheated steam and then passing through the catalyst holding tubes. The proper material of construction to use for the tubes holding the catalyst is

(A) Carbon steel
(B) Aluminum
(C) Austenitic steel
(D) Incoloy 800

Rate data and interpretation:

MB17: A reversible reaction of the type $A \rightleftarrows B$ was studied with C_{A0} = 0.8 lbmol/ft³ and C_{B0} = 0.5 lb mol/ft³. Equilibrium conversion X_{AE} is 0.35. The concentration data taken as a function of time are plotted in the figure below.

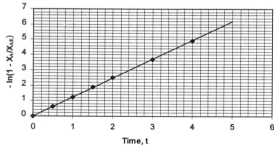

The forward reaction specific rate constant [h⁻¹] is nearer to

(A) 0.93   (B) 0.743   (C) 0.66   (D) 0.74

MB18: Specific rate constants were experimentally determined for a reaction and are given below

T °K	303	313	323	333	343
k $\frac{liters}{gmol.K}$	0.5	1.1	2.2	4.0	6.0

The activation energy [cal/gmol] of this reaction is near to

(A) 13072 (B) 20000 (C) 14800 (D) 18600

MB19: The first order liquid phase reversible reaction $A \underset{k_2}{\overset{k_1}{\rightleftarrows}} R$ takes place in a batch reactor. After 8 minutes, conversion is 33.3 % while the equilibrium conversion is 66.7 %. The specific rate constant [h-1] for the reverse reaction is close to

(A) 0.1 (B) 0.029 (C) 0.29 (D) 0.32

MB20: The gas phase reaction was carried out in a constant volume batch reactor. Times for 50 % conversion were determined at various concentrations and at a temperature of 110 °C. The data are reported as follows

C_{Ao}	ln C_{Ao}	$t_{1/2}$	$t_{1/2}$
0.01	-4.61	5	1.61
0.025	-3.69	2	0.69
0.05	-3	1	0
0.075	-2.59	0.67	-0.4
0.1	-2.3	0.5	-0.69

The order of the reaction is

(A) 3 (B) 2 (C) 1.5 (D) 1

Reactor types:

MB21 to MB22: A homogeneous liquid phase reaction

$$2A \rightarrow R \qquad -r_A = k C_A^2$$

is carried out in a mixed reactor. A 50 % conversion is attained.

MB21: If the reactor is replaced by another one 6 times as large and all other factors are kept the same, the percent conversion that will be attained is near to

(A) 67 (B) 91.7 (C) 95 (D) 75

MB22: If the mixed reactor is replaced by an equal size plug flow reactor, the conversion [%] attained will be nearer to

(A) 67 (B) 85 (C) 90 (D) 75

MB23 to MB24: The hydrolysis of acetic anhydride is carried out in a series of 4 same size mixed flow reactors. The reactors are operated at different temperatures 10, 15, 25 and 40 °C respectively. The inlet composition is 1.5 lb moles/gal and the flow rate is 30 gpm. The desired conversion is 95 %. The reaction is first order and the rate constants are

Temperature °C	10	15	25	40
k min^{-1}	0.0567	0.0806	0.158	0.38

MB23: The size of the reactor [gpm] required to effect 95 % conversion is nearer to

(A) 190   (B) 200   (C) 230   (D) 175

MB24: If all the reactors are maintained at a temperature of 15 °C, the number of reactors each of the same size as in problem KI2-3 is

(A) 5   (B) 4   (C) 3   (D) 7

PLANT DESIGN

Economics:

MB25 to MB28: It is contemplated to install a distillation system as shown in figure PA2-1. An estimate of capital investment in equipment and annual variable costs base on 1998 year cost index is to be made. The available cost data are as follows

Cost of tower Shell and internals		cost of vessels		cost of heat exchangers		cost of pumps (centrifugal)	
Tower Diameter In	cost $ per tray	Cap gal	cost $	A, ft^2	cost $	gpm x psi	cost $
60	1200	100	1000	1000	17250	3000	7200
70	1500	200	1600	1400	21150	4000	8000
80	1850	300	1900	1800	24600	5000	8600
90	2250	400	2200	2200	27750	10000	11600
100	2700	500	2600	2600	30300	15000	16000

(a) All costs are installed costs of equipment.

(b) All costs are 1979 costs. 1979 chem. engineering cost index = 230
    1998 chem. engineering cost index = 390

(c) Fixed capital costs comprise of installed equipment cost + 60 % of installed equipment cost for piping, instrumentation and other auxiliaries.

(d) Annual variable cost equals 15 % of fixed capital cost.

(e) All costs are based on carbon steel as material of construction.

(f) Operating hours per year = 8640 h

Figure PA2-1

C101	HE101	HE102	V101		
Distillation Column	Overhead condenser	Reboiler	Reflux drum		
No. of valve trays = 30	Q = 9.5 MMBtu/h	Q = 11.22 MMBtu/h	3'-0" Dia x 4'0" T-T		
8'- 0" DIA x75' 0"T-T	A = 1270 ft^2	A = 2500 ft^2	Cap. 210 gal		
Spacing 24"	M/C - CS Shell and tubes	M/C - CS Shell and tubes	M/C - CS		
M/C-CS shell & trays					

P101	P102
Reflux pump	Bottoms Pump
350 gpm x 35 psi	50 gpm x 50 psi
Type - centrifugal	Type - centrifugal
M/c - CS	M/c - CS

Other data are as follows

Equipment	Size basis	Cost exponent	Lang factor
Pump	HP	0.52	4.0
Shell & tube Exchanger	Area, A	0.62	4 (C.S.) 3.3 (SS)
Column	Dia.	0.65	4
Atm. storage tanks	gal	0.60	4
Pressure vessel	gal	0.60	4 (CS) 2 (SS)

MB25: Total fixed capital investment ($) for the distillation column is near to

    (A) 128000 (B) 205000   (C) 150000 (D) 168000

MB26: If cooling water cost is $0.1/1000 gal, the annual cost ($) for cooling water is close to

(A) 12700   (B) 19500   (C) 24600   (D) 36000

MB27: If the cost of steam is $2/1000 lb of steam, the annual cost ($) for steam is near to

(A) 226,200   (B) 239,000   (C) 220,000   (D) 235,000

MB28: Annual variable cost ($) for the entire installation is near to

(A) 226200   (B) 185,300   (C) 200,000   (D) 300,000

THERMODYNAMICS

Compressors and expanders:

MB29 to MB30: A Steam turbine operates adiabatically under conditions as shown in the following figure. The steam flow rate is 20,000 lb/h. Steam exhausts at 20 psia.

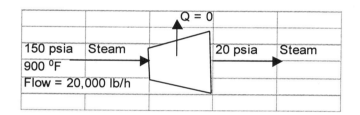

MB29: If the turbine efficiency is 85 %, the work output of the turbine [hp] is near to

(A) 1600   (B) 2200   (C) 2000   (D) 1800

MB30: If the steam is throttled to 100 psia and turbine efficiency declines to 80 %, the steam exit temperature [°F] would be near to

(A) 500   (B) 600   (C) 565   (D) 635

MB31 to MB32: Ammonia is to be compressed from 14 psia and 50 °F to a pressure of 60 psia in a single stage. The properties of ammonia at the inlet of the compressor are given in the diagram below

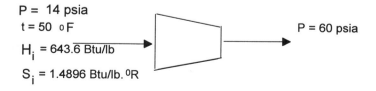

The properties of ammonia at 60 psia can be calculated from the following table.

	60 psia		
temp °F	v ft³/lb	H Btu/lb	S Btu/lb.°F
220	7.019	728.6	1.4658
240	7.238	739.7	1.4819
260	7.457	750.9	1.4976
280	7.675	762.1	1.5130
300	7.892	773.3	1.5281

Above 300 °F, specific heat of ammonia can be assumed constant and is 0.6 Btu/lb.°F.

MB31: If the compressor efficiency is 80 %, actual work [Btu/lb of ammonia] done by the compressor is
    (A) -105.8   (B) +105.8   (C) -102   (D) -132

MB32: Entropy increase [Btu/lb.°R] in the compression process is near to
    (A) 0.4896   (B) 0.0417   (C) 0.00315   (D) 0.217

Estimation of thermodynamic properties:

MB33: The following data are available in respect of benzyl acetate.

$P_C = 31.4$ atm, $T_C = 699$ K, normal BP = 213.5 °C
Heat of vaporization at normal BP = 11035 cal/g mol

The entropy change [Btu/lb.R] when benzyl acetate is vaporized to saturated vapor from its saturated liquid state at 25 °C is near to
    (A) 0.157 (B) 0.314 (C) 0.183 (D) 0.266

MB34: Enthalpy and entropy values for methyl chloride at 240 °F are given at two pressures in the following table.

Properties of methyl chloride at 240 °F

P (pressure) psia	V (specific volume) ft³/lb	H (enthalpy) Btu/lb	S (entropy) Btu/lb.R
6	24.71	246.6	0.5607
200	0.668	237.12	0.4129

The fugacity coefficient of methyl chloride at 240 °F and 200 psia is
    (A) 0.91 (B) 0.95 (C) 0.80 (D) 0.84

MB35: One method to calculate residual entropy $\Delta S'$ from PVT data is to express $\Delta S'$ in terms of partial derivative of residual volume as given below

$$\Delta S' = -\int_0^P \left(\frac{\delta \Delta V}{\delta T}\right)_P dP \qquad \text{(Constant T)}$$

The following data are available for isobutane vapor.
    Ideal gas molar specific heat :

$$C_p = 3.332 + 41.786 \times 10^{-3} T - 7.325 \times 10^{-6} T^2$$

where T is in degrees Rankine and C_P is in Btu/lbmol.R

Reference state (to be arbitrarily set): $H_0' = 150$ Btu/lb and $S_0' = 0.323$ Btu/lb.R

From PVT data for isobutane vapor, the integral $\int_0^P \left(\frac{\delta \Delta V}{\delta T}\right)_P dP$ was determined upto a pressure of 228.3 psia and was found to be -0.103 psia.ft²/lb.R. With these data, the entropy [Btu/lb.R] of isobutane vapor at 190 °F and 228.3 psia is near to
    (A) 0.405 (B) 0.312 (C) 0.296 (D) 0.285

MB36: Specific volume of superheated SO_2 at 1000 psia and 480 °F is 0.1296 ft³/lb. Its residual volume [ft³/lb] at these conditions is

    (A) 0.041 (B) 0.028 (C) 0.062 (D) 0.057

V-L Composition of miscible and partially miscible systems:

MB37 to MB40: i-Butanol forms a minimum boiling-point azeotrope. The t-x diagram for this system is shown in the following figure.

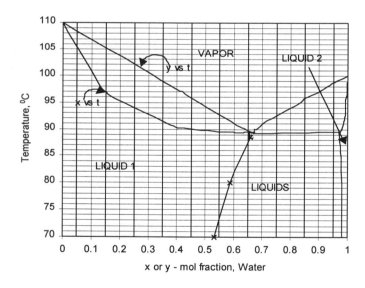

i-Butanol - Water system

MB37: i-Butanol concentration [wt %] of the azeotrope is most nearly
    (A) 61.7 (B) 66.04 (C) 69.1 (D) 56.3

MB38: An aqueous mixture of i-butanol and water contains 95 wt % water. The bubble point [°C] of this liquid is most nearly

(A) 94   (B) 88.5   (C) 100   (D) 97

MB39: 10,000 lb of a vapor mixture of 23 mol % i-butanol and 77 mol % water vapor at 105 °C is condensed at a constant pressure of 1 atm. and subcooled to a temperature of 80 °C. After phase separation, the butanol concentration [wt %] of butanol rich layer will most nearly be

(A) 0.8066   (B) 0.95   (C) 0.7562   (D) 0.44

MB40: A vapor mixture containing 50 mol % i-butanol and 50 mol % water vapor at 110 °C and 1 atm pressure will have a dew point temperature most nearly

(A) 95   (B) 100   (C) 92.8   (D) 108

STOP. CHECK YOUR WORK. END OF PM EXAM 3..

AFTERNOON SAMPLE EXAMINATION

Instructions for afternoon Session

1. You have four hours to work on the afternoon session. Do not write in this handbook.

2. Answer all forty questions for a total of forty answers. There is no penalty for guessing.

3. Work rapidly and use your time effectively. If you do not know the correct answer, skip it and return to it later.

4. Some problems are presented in both metric and English units. Solve either problem.

5. Mark your answer sheet carefully. Fill in the answer space completely. No marks on the workbook will be evaluated. Multiple answers receive no credit. If you make a mistake, erase completely.

Work all 40 problems in four hours.

P.E. Chemical Engineering Exam
Afternoon Session

Ⓐ Ⓑ Ⓒ Fill in the circle that matches your exam booklet.

AB1 Ⓐ Ⓑ Ⓒ Ⓓ	AB11 Ⓐ Ⓑ Ⓒ Ⓓ	AB21 Ⓐ Ⓑ Ⓒ Ⓓ	AB31 Ⓐ Ⓑ Ⓒ Ⓓ
AB2 Ⓐ Ⓑ Ⓒ Ⓓ	AB12 Ⓐ Ⓑ Ⓒ Ⓓ	AB22 Ⓐ Ⓑ Ⓒ Ⓓ	AB32 Ⓐ Ⓑ Ⓒ Ⓓ
AB3 Ⓐ Ⓑ Ⓒ Ⓓ	AB13 Ⓐ Ⓑ Ⓒ Ⓓ	AB23 Ⓐ Ⓑ Ⓒ Ⓓ	AB33 Ⓐ Ⓑ Ⓒ Ⓓ
AB4 Ⓐ Ⓑ Ⓒ Ⓓ	AB14 Ⓐ Ⓑ Ⓒ Ⓓ	AB24 Ⓐ Ⓑ Ⓒ Ⓓ	AB34 Ⓐ Ⓑ Ⓒ Ⓓ
AB5 Ⓐ Ⓑ Ⓒ Ⓓ	AB15 Ⓐ Ⓑ Ⓒ Ⓓ	AB25 Ⓐ Ⓑ Ⓒ Ⓓ	AB35 Ⓐ Ⓑ Ⓒ Ⓓ
AB6 Ⓐ Ⓑ Ⓒ Ⓓ	AB16 Ⓐ Ⓑ Ⓒ Ⓓ	AB26 Ⓐ Ⓑ Ⓒ Ⓓ	AB36 Ⓐ Ⓑ Ⓒ Ⓓ
AB7 Ⓐ Ⓑ Ⓒ Ⓓ	AB17 Ⓐ Ⓑ Ⓒ Ⓓ	AB27 Ⓐ Ⓑ Ⓒ Ⓓ	AB37 Ⓐ Ⓑ Ⓒ Ⓓ
AB8 Ⓐ Ⓑ Ⓒ Ⓓ	AB18 Ⓐ Ⓑ Ⓒ Ⓓ	AB28 Ⓐ Ⓑ Ⓒ Ⓓ	AB38 Ⓐ Ⓑ Ⓒ Ⓓ
AB9 Ⓐ Ⓑ Ⓒ Ⓓ	AB19 Ⓐ Ⓑ Ⓒ Ⓓ	AB29 Ⓐ Ⓑ Ⓒ Ⓓ	AB39 Ⓐ Ⓑ Ⓒ Ⓓ
AB10 Ⓐ Ⓑ Ⓒ Ⓓ	AB20 Ⓐ Ⓑ Ⓒ Ⓓ	AB30 Ⓐ Ⓑ Ⓒ Ⓓ	AB40 Ⓐ Ⓑ Ⓒ Ⓓ

PM SAMPLE EXAM 4
FLUIDS
Properties of fluids:

AB1: The value of ω, the acentric factor is required in many correlations developed for the calculation of physical properties of liquids and gases. A few relations are available to calculate the value of ω. One such relation is as follows

$$\omega = -\log_{10} P_r^{sat} - 1.0$$

where P_r^{sat} is the reduced vapor pressure at $T_r = T/T_c = 0.7$

The following data are available for toluene in Perry's Handbook.

$t_c = 320\ ^\circ C,\ P_c = 41.6$ atm.,
Vapor pressures of toluene

P atm.	1	2	3	5	10
Temp °C	110.6	136.5	178	215.8	262.5

The acentric factor for toluene is nearer to

(A) 0.35 (B) 0.25 (C) 0.4 (D) 0.29

AB2: Watson's correlation allows to calculate the heat of vaporization of a liquid at another temperature if the heat of vaporization is available at one temperature. Heat of vaporization of propyl alcohol at its boiling point ($t_b = 97.2\ ^\circ C$) is 164.36 cal/g. Its critical temperature is 263.7 °C. Its heat of vaporization [J/gmol.] at 115 °C is

(A) 27460 (B) 36840 (C) 47573 (D) 39625

AB3: 10 lbs of water at 70 °F in a container is heated to 1500 °F and 260 psia. The enthalpy change [Btu] in this process is

(A) 1567.7 (B) 15677.6 (C) 12762 (D) 17543

AB4: Jamieson and Tudhope recommended the following relation for the estimation of thermal conductivity of a solution of electrolytes.

$$k_{mix}(20\ ^0C) = k_{H_2o}(20\ ^0C) + \frac{1}{4.186} \Sigma \sigma_i C_i$$

where k_{mix} (20 °C) = Thermal conductivity of electrolyte solution at 20 °C, cal/(cm.s.K)
$k_{H_2o}(20^0C)$ = Thermal conductivity of water at 20 °C, cal/(cm.s.K)
C_i = Concentration of electrolyte in solution, gmol/L
σ_i = a coefficient which is characteristic of each ion.

Thermal conductivity of a solution at another temperature can then be calculated using the relation given below

$$k_{mix}(T) = k_{mix}(^0C)\frac{k_{H_2o}(T)}{k_{H_2o}(20\ ^0C)}$$

The following pertinent data are available

k for water at 20 °C = 0.340 Btu/h.ft.°F
k for water at 20 °C = 0.381 Btu/h.ft.°F
σ_i for Na⁺ = 0000, σ_i for OH^- = 20.934x10⁻⁵

The thermal conductivity [Btu/(h.ft.°F] of 10 % NaOH solution at 60 °C will approximately be

(A) 0.418   (B) 0.374   (C) 0.352   (D) 0.321

KINETICS

Control of reactors:

AB5: When an initial batch heat up is required, the reaction temperature control is best accomplished by
  (A) Cascade temperature control
  (B) Split range cascade control
  (C) Emergency control
  (D) None of the above

AB6: The overall transfer function for two CSTRs in series with feed back controller is represented by the following block diagram.

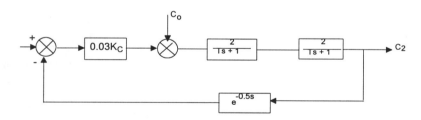

If a unit step function is applied to the set point, the offset is near to

(A) $\frac{1}{1+0.12K_C}$ (B) $\frac{0.12K_C}{1+0.12K_C}$ (C) $0.12K_C$ (D) 0

AB7 to AB8: The calculated data on heat generation and removal for a continuous CSTR carrying out a reaction with first order kinetics are plotted in the following figure. The heat removal data are calculated with a jacket temperature of 113 °F.

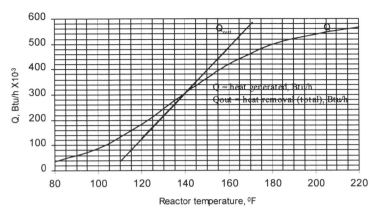

Heat generation and removal data

The design conditions for the reactor are as follows

Feed = 2000 lb/h at 70 °F, C_{ao} = 0.5
Hold up = 1.2 h
C_P = 0.75 Btu/(lb.°F)
Jacket area, A = 50 ft^2
U = Overall heat transfer coefficient, Btu/(h.ft^2.°F)
ΔH = heat of reaction = - 600 Btu/lb
First order reaction: k = 1.3 h^{-1} at 150 °F, the reaction rate constant.
E = 30000 Btu/lbmol, (activation energy)

AB7: At 50% conversion, the critical temperature difference ΔT_c, [°F] is near to

(A) 20.1   (B) 26.2   (C) 23.8   (D) 32.8

AB8: If the reaction is carried out at steady state point represented by the intersection of the heat and removal curves, the following can be said about the reactor stability

(A) The reactor operation is stable.
(B) The reactor operation is stable because the jacket temperature is constant.
(C) The reactor operation will be stable if the feed temperature is reduced by 50 °F.
(D) The reactor can be operated at this steady state by using a feed back control
    to regulate the jacket temperature.

[Hints:

(1) Heat generated = Q = kVc$_o$(1-x)($-\Delta H_R$) = $Fc_o(-\Delta H_R)x$

(2) $\dfrac{x}{1-x} = \dfrac{kV}{F}$

(3) $k = ae^{-\frac{E}{RT}}$

(4) $Q_{out} = UA(T - T_j) + F\rho C_P(T - T_F)$

(5) For first order reaction, $\left(\frac{\partial Q}{\partial T}\right)_x = \left(\frac{dQ}{dT}\right)_{ss}\left(\frac{1}{1-x}\right)$

(6) Critical temperature difference = $\Delta T_c = \frac{RT^2}{E}$

For stability, $\quad UA + F\rho C_P \frac{2-x}{1-x} > \left(\frac{\partial Q}{\partial T}\right)x$

$$UA + F\rho C_p > \left(\frac{\partial Q}{\partial T}\right)_x (1-x)$$

$$T - T_j < \Delta T_c$$

Q = Heat generated due to reaction, Btu/h

Q_{out} = Heat leaving the reactor through jacket, Btu/h

x = Conversion $\quad\quad$ k = Reaction rate constant

T = Reactor temperature, °F $\quad$ T_j = Jacket wall temperature

ΔT_c = *Critical temperature* difference

E = Activation energy, Btu/lbmol, V = volume of reactor, ft³

F = Feed, lb/h $\quad$ *ρ = Density of solution*, lb/ft³ $\quad$ C_P = specific heat Btu/(lb.°F)

U = Heat transfer coefficient, Btu/h.ft².°F $\quad$ A = Jacket heat transfer surface. $\quad$]

Endothermic/exothermic reactors:

AB9 to AB11: The following aqueous phase reaction is carried in a mixed reactor as shown in the figure

$$A \underset{k_2}{\overset{k_1}{\rightleftarrows}} R$$

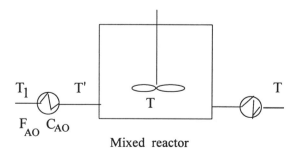

Mixed reactor

The feed to the reactor is first cooled, and then fed to the reactor and the product withdrawn from the reactor is cooled to 25 0C before sending to storage. The reaction data and other property data are as follows

Feed rate = 980 gmol/min, product free

C_{A0} = 4 gmol/liter   Rate constants:

$$k_1 = e^{17.2 - \frac{11600}{RT}} \qquad k_2 = e^{41.9 - \frac{29600}{RT}}$$

where T is in degrees kelvin.

Heat of reaction ΔH_r = -18000 cal/gmol of A

$\Delta C_p = 0$ and $C_p = C_p$ of water

Pressure of reaction is constant and equal to 1 atm.

The locus of maximum conversion vs temperature is shown in Figure

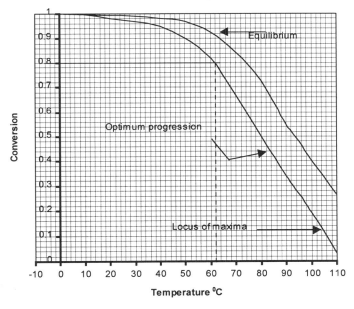

Equilibrium and optimal progression conversions as a function of temperature

AB9: Volume (liters) of a mixed flow reactor is near to
(A) 2500   (B) 2000   (C) 2750   (D) 1750

AB10: The slope [K^{-1}] of the energy balance line is close to
(A) 1/72   (B) 1/36   (C) 1/54   (D) 1/60

AB11: Total cooling duty (kcal/min) is near to

(A) 13.965x10³   (B) 9.065x10³   (C) 15x10³   (D) 12.2 x10³

AB12: Optimal rates of a reaction obtained by using the locus of maximum conversions at various temperatures are given below as a function of the conversion, X_A.

conversion	0.3	0.5	0.6	0.7	0.8
$-r_A$	5.8	2.4	1.47	0.83	0.4

If a plug flow reactor is used, the minimum space-time [min] for 80 % conversion is near to

      (A) 1.66   (B) 1.52   (C) 3.8   (D) 1.83

Product distribution:

AB13: Two liquid phase parallel reactions, of which the first one is desirable proceed as follows

$$A+B \xrightarrow{k_1} R+T \qquad -r_{A_1} = k_1 C_A C_B^{0.2}$$
$$A+B \xrightarrow{k_2} S+U \qquad -r_{A_2} = k_2 C_A^{0.5} C_B^{1.7}$$

From the point of product distribution, the most appropriate contacting scheme is

      (A) C_A high, C_B low
      (B) C_A low, C_B low
      (C) C_A high, C_B high
      (D) C_A low, C_B high

AB14 to AB15: For consecutive reactions $A \xrightarrow{k_1} B \xrightarrow{k_2} C$, the values of rate constants k_1 and k_2 are

$k_1 = 0.35$ h^{-1} $k_2 = 0.13$ h^{-1} $C_{A0} = 5$ lb moles/ft^3, $C_{B0} = 0$, $C_c = 0$ at $t = 0$

The rate equations for the two reactions and their solutions in case of a batch reactor are

$$r_a = -\frac{dC}{dt} = k_1 C_A$$

$$r_b = -\frac{dC_B}{dt} = k_2 C_B - k_1 C_a$$

$$C_A = C_{A0} e^{-k_1 t}$$

$$C_B = C_{A0} \frac{k_1}{k_2 - k_1} (e^{-k_1 t} - e^{-k_2 t})$$

$$C_c = C_{A0} \left(1 - \frac{k_2}{k_2 - k_1} e^{-k_1 t} + \frac{k_1}{k_2 - k_1} e^{-k_2 t}\right)$$

For a continuous mixed reactor, the following relations apply

$$C_{A1} = \frac{C_{A0}}{1 + k_1 \theta}$$

$$C_{B1} = \frac{k_1 C_{A0}\theta}{(1+k_1\theta)(1+k_2\theta)}$$

where θ = residence time in the vessel

AB14: The maximum concentration [lb mole/ft³]attained by B when operating as a batch reactor is near to

(A) 2.64   (B) 2.53   (C) 2.79   (D) 2.94

AB15: For a single stage continuous mixed reactor, the maximum concentration [lb mol/ft3] of B in the effluent stream is close to

(A) 1.62   (B) 1.43   (C) 1.925   (D) 1.74

AB16: A decomposes to products B, C, and D according to the following reactions

$$A \xrightarrow{k_1} B \qquad A \xrightarrow{k_2} C \qquad A \xrightarrow{k_3} D$$

The values of the reaction specific rates are $k_1 = 0.36$ h⁻¹ $k_2 = 0.12$ h⁻¹ $k_3 = 0.1$ h⁻¹
At start of the reaction, $N_{ao} = 100$ gmol, $N_{bo} = N_{co} = N_{do} = 0$. After one hour of reaction, the gmols of D in the reaction mixture will be

(A) 7.59   (B) 9.1   (C) 22.8   (D) 17.1

Reactor design:

AB17 to AB20: A solution containing a reactive component (Initial concentration, $C_{ao} = 0.5$ lb mole/ft³) is to be treated in different types of reactors. The feed rate for continuous flow operation is to be 25 ft³/h. The reaction rate data for the decomposition of A are as follows

C_A, lb mole/ft3	-r_A, lb moles/(h.ft³)	-$1/r_A$
0.5	0.85	1.1765
0.4	0.53	1.8868
0.3	0.31	3.2258
0.2	0.18	5.5556
0.1	0.081	12.3457
0.05	0.04	25.00

AB17: If the filling and draining time per batch is negligible, the number of batches that can be processed per day in a batch reactor is

(A) 12   (B) 20   (C) 15   (D) 10

AB18: If the reaction is carried out in a continuous mixed flow reactor and the processing rate is 25 ft³, volume [ft³] of the reactor required to realize 90 % conversion is near to

(A) 100.4   (B) 281.3   (C) 220.8   (D) 320.6

AB19: If each vessel has a volume of 50 ft³, the percentage conversion [%] attained with a 2 continuous mixed flow reactors is nearer to
(A) 90.2   (B) 75.4   (C) 85.2   (D) 95.1

AB20: If a plug flow reactor is used, the volume [ft³] of the reactor to effect 90 % conversion is close to
(A) 60.0   (B) 100   (C) 90   (D) 80

MASS TRANSFER

Phase diagrams:

AB21 to AB24: 500 lb of a mixture of acetone and water were mixed together with 500 lb of solvent trichloroethane and the mixture allowed to separate into two cojugate phases and come to equilibrium. The raffinate layer was analyzed and was found to have the following analysis

Component	Mass fraction in raffinate
Acetone	0.15
Trichloroethane	0.007
Water	0.843

The ternary phase equilibrium diagram together with conjugate line for the system is provided below:

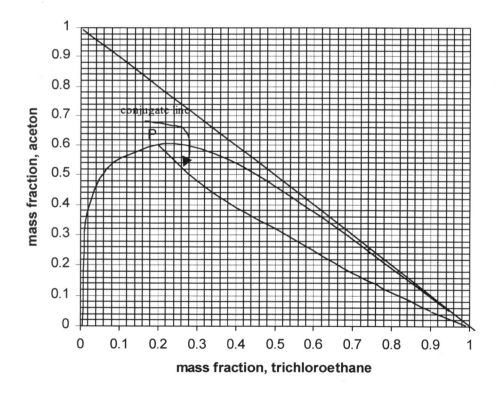

Phase diagram for acetone-water-trichloroethane system

AB21: The mass fraction of acetone in extract layer is near to
        (A) 0.54   (B) 0.254   (C) 0.3   (D) 0.18

AB22: The amount [lb/500 lb batch] of extract that is obtained is near to
        (A) 672.5   (B) 416.1   (C) 578.8   (D) 603.2

AB23: The solute mass fraction of acetone in the original mixture of acetone and water is near to
        (A) 0.5   (B) 0.3   (C) 0.44   (D) 0.38

AB24: The selectivity of the solvent trichloroethane for acetone is near to
        (A) 287   (B) 238   (C) 162.5   (D) 184

THERMODYNAMICS

Combustion:

AB25 to AB28: A municipality is considering installing a plant to burn its solid waste (garbage) in an incinerator and recover energy together with recovery of metals from it. The wet refuse after shredding and separation of metals and glass has the following ultimate analysis:

Component	H_2O	C	H	O	inerts
wt %	28	23.4	3.0	20	25.6

The following sketch shows the scheme for recovery of energy in the form of steam from the treated waste.

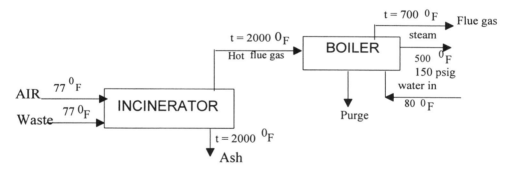

Complete combustion of carbon and hydrogen in the waste may be assumed. Other data are

Standard heats of formation:

Component	ΔH_f (kcal/gmol)
CO_2	-94.05
$H_2O(l)$	-68.314
$H_2O(v)$	-57.8

Molar specific heats* -Btu/(lb mol.°F)
temperature, °F

	700	2000
CO_2	10.262	12.01
H_2O (v)	8.341	9.339
N_2	7.055	7.542
O_2	7.347	7.976

* Average specific heat value between t = 0 °F and the temperature
  Specific heat of ash = 0.3 Btu/lb. °F

Very small amounts of sulfur and chlorine in the waste may be neglected in these calculations. Heat losses may also be ignored.

AB25: Based on available C and hydrogen in the waste, standard heat of combustion of the waste [Btu/lb] is near to

        (A) - 8500   (B) -9000   (C) -7060   (D) -5145

AB26: Percent excess air used is near to

        (A) 160   (B) 150   (C) 100   (D) 120

AB27: Total flue gas [lb mols/lb of waste] is near to

        (A) 0.306   (B) 0.402   (C) 0.286   (D) 0.203

AB28: The amount of steam [lb/lb garbage burned] is near to

        (A) 2.53   (B) 3.12   (C) 2.22   (D) 2.84

Economics:

AB29: Insulation is being considered for a new exchanger. the insulation is available in 1, 2, 3, and 4 in thicnesses. The following data have been determined for the different thicknesses.

Insulation thickness in	1	2	3	4
Cost of installed insulation $	1,440	1,920	2,160	2,244
Annual fixed charges $	173	230	259	269
Annual Savings due to less heat loss $	1,037	1,210	1,279	1,313

Annual minimum return of 15% is required. The choice for the thickness [in] of the insulation is

        (A) 1   (B) 2   (C) 3   (D) 4

AB30: An investment is expected to yield profits in the next 5 years of its useful life as follows

year	1	2	3	4	5
Annual cash flow to project after taxes,$	36000	37200	43200	48000	51600

The total fixed capital is $120,000, salvage value is $12,000 and working capital is $12,000. Assuming straight line depreciation and an interest rate of 15%, the time average value of net profits based on future worth at the end of project's useful life is nearer to

        (A) 21,600   (B) 37,232   (C) 20,443   (D) 23,356

AB31: A project is being considered with 3 alternate investment proposals. The data gathered for comparing these investments are as in the following table:

Investment Number	Total initial fixed-capital Investment $	Working capital $	Salvage value at end of useful life $	Service Life years	Annual cash flow to project after taxes $	Annual cash expenses $
1	110,000	11,000	11,000	5	39,600	48,000
2	187,000	11,000	17,000	7	57,200	30,800
3	231,000	16,500	22,000	8	65,000	23,100

On the basis of payout periods with no interest charge, the investment to be recommended is

(A) number 1   (B) number 2   (C) number 3   (D) None of three above.

AB32: The annual direct production costs for a plant operating at 70 % capacity are $350,000. Total annual fixed charges, overhead costs and general expenses are $250,000. If total sales are $700,000 and the product sells at $50 per unit, the break-even point in units of production is nearer to

(A) 12000   (B) 9000   (C) 10000   (D) 11000

Thermodynamic charts and data bases:

AB33: A refrigeration vapor compression uses HFC-134a as refrigerant. The condenser operates at 150 psia and the refrigerant leaves the condenser as saturated liquid. The evaporator temperature is 0 °F. The refrigeration cycle uses an adiabatic turbine with 85 % efficiency. The entropy increase [Btu/(lb°R)] during turbine expansion is

(A) 0.013   (B) 0.006   (C) 0.00   (D) 0.004

AB34: A 20 % NaOH solution at 70 °F is is mixed with a 70 % NaOH solution at 250 °F in equal amounts to produce a 45 % NaOH solution. The resulting solution is cooled to 70 °F. The heat to be removed [Btu/lb of solution] to effect this cooling is

(A) -115   (B) 115   (C) 120   (D) -132

AB35 to AB36: Ethane (critical constants : P_C = 48.2 atm., T_C = 305.4 K) is compressed from 90 °F and 15 atm. to 96 atm. and 145 °F in a steady flow process. The molar specific heat of ethane at zero pressure is given by the following equation

$$C_{p0} = 4.01 + 0.01636T$$ where T is in degrees Rankine.

Using the generalized enthalpy and entropy charts,

AB35: The work of compression [Btu/lb of ethane] will be

(A) -1859   (B) -62   (C) -110   (D) -132

AB36: The entropy change [Btu/(lb°R)] will be

      (A) 1.667   (B) 0.667   (C) 0.0523   (D) 0.112

Thermodynamic laws:

AB37: 10 mols/s of oxygen at 100 °C and 10 mols/s of nitrogen at 150 °C are continuously mixed in a vessel in a steady flow process. The heat loss from the vessel is given by $Q_L = 209(t-25)$ where t is in °C and Q_L in J/s. The specific heats of oxygen and nitrogen as a function of temperature are given below:

    O_2 $C_P = 6.9963 + 3.1 \times 10^{-3} t$
    N_2 $C_P = 6.8413 + 1.25 \times 10^{-3}$

Where t = Temperature, °C.
    C_P = Specific heat, cal/(g mol)(°C)

Assuming no heat of mixing, the exit temperature [°C] of the mixed stream from the vessel will be

      (A) 81.32   (B) 102.4   (C) 90.2   (D) 55.8

AB38: The entropy change [Btu/lb mol.°R] that takes place when one lb mol of ethyl ether originally at 68 °F and 1 atmosphere is heated to a state of superheated vapor at 122 °F and 1 atm pressure is near to

      (A) 5.8   (B) 0.32   (C) 12.40   (D) 23.67

The properties of ethyl ether are as follows

    Specific heat of liquid = 0.521 Btu/(lb.°F)
    Boiling point at 1 atm. = 94.3 °F
    Molecular weight = 74.12
    Heat of vaporization = 151.06 Btu/lb
    specific heat of vapor = 0.44 Btu/(lb.°F)
    (assume specific heats to be constant)

AB39: 1 lb of a mixture of water and steam is contained in a rigid vessel at 150 psia. Heat is added to the vessel until the contents of the vessel reach the condition of 500 psia and 700 °F. Heat [Btu/lb] required to be added to the vessel to reach the final state is near to

      (A) 602.4   (B) 632.6   (C) 571.3   (D) 655.8

AB40: A Carnot heat pump is used for heating a building The outside air is at 30 °F and is the cold reservoir. The building is to be maintained at 75 °F. 200,000 Btu/h are required for the heating. The heat [Btu/h] taken from outside is near to

      (A) 183,173   (B) 200,000   (C) 100,000   (D) 167,000

STOP. CHECK YOUR WORK. END OF PM EXAM 4..

AFTERNOON BONUS EXAMINATION

Instructions for afternoon Session

1. You have 96 minutes to work on this bonus exam.. Do not write in this handbook.

2. Answer all 16 questions for a total of 16 answers. There is no penalty for guessing.

3. Work rapidly and use your time effectively. If you do not know the correct answer, skip it and return to it later.

4. Some problems are presented in both metric and English units. Solve either problem.

5. Mark your answer sheet carefully. Fill in the answer space completely. No marks on the workbook will be evaluated. Multiple answers receive no credit. If you make a mistake, erase completely.

Work all 16 problems in 96 minutes.

P.E. Chemical Engineering Exam
Bonus Session

Ⓐ Ⓑ Ⓒ Fill in the circle that matches your exam booklet.

B1 Ⓐ Ⓑ Ⓒ Ⓓ	B11 Ⓐ Ⓑ Ⓒ Ⓓ
B2 Ⓐ Ⓑ Ⓒ Ⓓ	B12 Ⓐ Ⓑ Ⓒ Ⓓ
B3 Ⓐ Ⓑ Ⓒ Ⓓ	B13 Ⓐ Ⓑ Ⓒ Ⓓ
B4 Ⓐ Ⓑ Ⓒ Ⓓ	B14 Ⓐ Ⓑ Ⓒ Ⓓ
B5 Ⓐ Ⓑ Ⓒ Ⓓ	B15 Ⓐ Ⓑ Ⓒ Ⓓ
B6 Ⓐ Ⓑ Ⓒ Ⓓ	B16 Ⓐ Ⓑ Ⓒ Ⓓ
B7 Ⓐ Ⓑ Ⓒ Ⓓ	
B8 Ⓐ Ⓑ Ⓒ Ⓓ	
B9 Ⓐ Ⓑ Ⓒ Ⓓ	
B10 Ⓐ Ⓑ Ⓒ Ⓓ	

PM BONUS EXAM..

Equipment design:

B1: A pump is delivering 800 gam of a liquid to a storage tank situated 600 ft away. The viscosity of the liquid is 0.31 centistokes and its density is 46.4 lb/ft³. The pump discharge pressure is 75 psia. Liquid temperature is 312 °F. Material of construction of piping is carbon steel. Based on usual criteria for line sizing, standard pipe size (schedule and diameter in inches) to be recommended is

        (A) Schedule 80, 6"
        (B) Schedule 40, 6"
        (C) Schedule 40, 4"
        (D) Schedule 40, 8"

THERMODYNAMICS

Combustion:

B2 to B3: Below are given some reactions together with the heats of formation (kcal/gmol) of the compounds involved in the reaction.

(1) $CH_3CHO(g) + H_2(g) \rightarrow C_2H_5OH(g)$
              -39.72   -52.23

(2) $C_2H_5OH(g) + 3O_2(g) \rightarrow 2CO_2(g) + 3H_2O(l)$
              -52.23   -94.052   -68.32

(3) $H_2(g) + \tfrac{1}{2}O_2(g) \rightarrow H_2O(l)$
                                        -68.32

(4) $H_2O(l) \rightarrow H_2O(g)$
              -68.314   -57.8

B2: From these data, the heat of combustion [kcal/gmol] of gaseous acetaldehyde at 25 °C and one atmosphere pressure is near to

        (A) -285.02   (B) -263.98   (C) -318.6   (D) -304.2

B3: The net heating value [Btu/ft³] measured at 60 °F and atmospheric pressure and saturated with water vapor is near to

        (A) 1000   (B) 1230   (C) 1120   (D) 1050

Economics:

B4: A rough rule of chemical industry is that $1 of annual sale requires $1 of fixed capital investment. In a chemical process plant, the total capital investment is 3 million dollars and the working capital is 20 % of the total capital investment. Annual total product cost is $180,000. The taxes are 48% of gross earnings.
The percent of total investment returned annually as gross earnings is close to

(A) 20   (B) 15   (C) 18   (D) 25

B5: A owes B the sum of $6000 due December 31, 1995 and $4000 on December 31, 1997. By mutual agreement, the loan was rescheduled for payment whereby A will make a payment of $8000 on December 31,1998 and a balance payment December 31, 1999 with interest at 10%. The amount of the final payment will be

(A) 4824.6   (B) 3800.18   (C) 4000.23   (D) 4200.67

Equilibrium data:

B6 to B7: The system ethyl alcohol (A) and toluene (B) forms an azeotrope at 76.65 °C and one atm. pressure. The composition of the azeotrope is 81 mol % ethyl alcohol. Using this composition and the fact that $y_i = x_i$ at azeotropic composition, the following values for Van Laar constants were obtained for the system.

$$A = 0.6596 \qquad B = 0.7371$$

The vapor pressures of the two components are given by following Antoine equations

Ethyl Alcohol: $\qquad \log P_1 = 8.3211 - \frac{1718.1}{t + 237.52}$

Toluene: $\qquad \log P_2 = 6.9301 - \frac{1344.8}{t + 221.4}$

where   t = temperature, °F   and P = vapor pressure, mm Hg.
Assume that the vapor behaves as an ideal gas.

B6: The boiling point [°C] of a mixture containing 60 mol% ethyl alcohol and 40% mol% toluene at 1 atm. pressure is near to

(A) 84.2   (B) 88.6   (C) 78.4   (D) 74.8

B7: The mol fraction of ethyl alcohol in the vapor in equilibrium with a liquid containing 60 mol% ethyl alcohol and 40 mol% toluene at a total pressure of 1 atm. is near to

(A) 0.745   (B) 0.86   (C) 0.692   (D) 0.66

Estimation of properties:

B8: Redlich and Kwong proposed a two constant equation of state to represent PVT behavior of real gases as follows

$$P = \frac{RT}{V-b} - \frac{a}{T^{\frac{1}{2}} V(V-b)} \quad \text{where} \quad a = \frac{0.4278\, R^2 T_c^{2.5}}{64 P_c} \quad \text{and} \quad b = \frac{0.0867 RT_c}{P_c}$$

An alternative form of RK equation obtained by multiplication of the above equation by V/RT is

$$Z = \frac{1}{1-h} - \frac{A}{B}\left(\frac{h}{1+h}\right) \quad \text{and} \quad h = \frac{b}{V} = \frac{BP}{Z}$$

where $B = b/RT$ and $A/B = a/bRT^{1.5}$. The constants A and B can be related to reduced pressure P_r and temperature T_r by the following relations

$$B = \frac{0.0867 P_r}{PT_r} \quad \text{and} \quad \frac{A}{B} = \frac{4.934}{T_r^{1.5}}$$

Substitution of these values of A and B in the modified RK equation yields

$$Z = \frac{1}{1-h} - \frac{4.934}{T_r^{1.5}} \frac{h}{1+h}$$

If V is known, it is more convenient to use $\quad h = \frac{b}{V} = \frac{0.0867 RT_c}{VP_c}$

Critical constants for propane are $P_c = 42$ atm. $T_c = 96.8\ ^\circ C$. 50 lb of propane are stored in 5 ft³ container at a temperature of 212 °F. The pressure [psig] developed by propane under these conditions with the use of RK equation will be

(A) 636.3 (B) 708.4 (C) 665.2 (D) 685.6

B9: Ethylene at 500 atm. and a temperature of 100 °C is contained in a cylinder of internal volume 2 ft³. The pounds of ethylene contained in the cylinder is near to

(A) 65.4 (B) 60.4 (C) 68.6 (D) 62.8

[Hint: Use generalized compressibility chart.]

Estimation of thermodynamic properties:

B10: Good estimations of the heats of vaporization of liquids are obtained if one uses the reduced form of Classius-Clapeyron equation given by

$$\frac{\Delta H_v}{ZRT_c} = \frac{B}{T_c}\left[\frac{T_r}{T_r + C/T_c}\right]^2$$

where B and C are two of the constants of the Antoine equation,
The critical constants for benzyl acetate are: $P_C = 31.4$ atm., $T_C = 699$ K, and its vapor pressures are given by Antoine equation

$$\ln P = A - \frac{B}{T + C}$$

where T is in K, and the constants are A = 16.5956, B = 4104.84, and C = −74.56. Under these conditions, the heat of vaporization [Btu/lb] of benzyl acetate at its normal boiling point is

(A) 136   (B) 132.2   (C) 142   (D) 123

B11: The enthalpy of 1-butene at 1000 psia 400 °F relative to the state of saturated liquid at 32 °F is 8560 cal/g mol. The critical constants for 1-butene are: P_C = 39.7 atm., T_C = 147 °C, and ω = 0.187. Its internal energy U, [Btu/lb] at the same pressure and temperature is

(A) 126   (B) 131   (C) 96.6   (D) 128

Thermodynamic charts and data bases:

B12: A vessel contains 15000 lb of ethylene at a pressure of 1050 psig. and 75 °F. Two thirds of this amount is removed as feed to another process. The temperature of ethylene in the vessel remains constant at 24 °C throughout the withdrawal. At the end of the withdrawal, the pressure guage on the vessel will indicate a pressure [psig] of

(A) 660   (B) 620   (C) 630   (D) 645

B13: Mollier diagram of steam allows to read enthalpy, entropy and other properties of steam to follow the processes, steam undergoes. Superheated steam originally at 200 psia 400 °F expands through a nozzle to a pressure of 25 psia. Using Mollier diagram for steam, the quality of steam at nozzle exit can be found as

(A) 83.3   (B) 88.7   (C) 95.2   (D) 80.6

V-L Composition of miscible and partially miscible systems:

B14 to B15: A batch of Nitrobenzene which is essentially insoluble in water is to be purified from a heavier impurity. It is charged to a steam heated kettle and water is added to it continuously to maintain the water level during distillation. The impurity does not affect the vapor pressure of nitrobenzene. The vapor pressure data for nitrobenzene and water are given below

temperature °C	Nitrobenzene vapor pressure (mmHg)	Water vapor pressure (mmHg)
60	2.72	149
70	4.8	233
80	8.13	355
90	13.3	526
95		634
100	21.0	760

Vapor pressures of water can be more accurately calculated by the following Antoine equation

$$\ln P = 18.3036 - \frac{3816.44}{t + 227.02}$$

where t = temperature, °F, and P = vapor pressure, mm Hg

B14: The boiling point [°C] of the mixture of nitrobenzene and water at atm. pressure is

(A) 99.3   (B) 100.0   (C) 95   (D) 98.7

B15: Mol fraction of nitrobenzene in vapor is

(A) 0.276   (B) 0.02676   (C) 0.54   (D) 0.054

Thermodynamic laws:

B16: A Carnot engine which receives heat at a certain temperature, develops 2 hp, and rejects 7500 Btu/h to a sink at 60 °F. The heat source temperature [°F] of the Carnot engine is near to

(A) 400   (B) 450   (C) 413   (D) 520

Stop. Check your work. End of Bonus Exam.

SOLUTIONS TO PM SAMPLE EXAM 1
FLUIDS
Bernoulli's equation:

MA1:

$$\frac{P_A}{\rho_A} + Z_A + \frac{U_A^2}{2g_c} - F + w = \frac{P_B}{\rho_B} + Z_B + \frac{U_B^2}{2g_c}$$

w = 0 U_A = 0 (large cross section of tank)

F = 28.5 ft given. $\rho_A = \rho_B$ (incompressible fluid)

$$U_B = \frac{101.2}{7.48 \times 60 \times 0.0233} = 9.7 \text{ ft/s} \qquad \frac{U_B^2}{2g_c} = \frac{9.7^2}{64.4} = 1.46 \cong 1.5 \text{ ft}$$

Bernoulli equation reduces to $Z_A - Z_B = F + \frac{U_B^2}{2g_c} = 28.5 + 1.5 = 30$ ft

Height from the bottom of tank = 30 - 10 = **20 ft**

Answer is (C)

MA2:

$$\frac{P_A}{\rho_A} + Z_A + \frac{U_A^2}{2g_c} - F + w = \frac{P_B}{\rho_B} + Z_B + \frac{U_B^2}{2g_c}$$

Bernoulli equation reduces to: $w = F + (Z_B - Z_A) + \frac{U_B^2}{2g_c}$

Velocity through 2″ pipe = $3(3.068/2.067)^2$ = 6.61 ft/s

$$\frac{U_B^2}{2g_c} = \frac{6.61^2}{64.4} = 0.68 \text{ ft}$$

∴ w = 12 + (40 – 10) + 0.68 = 42.68 ft

Pressure developed = $\frac{42.68 \times 1.605 \times 62.4}{144} = 29.7$ psia

Therefore, pressure indicated by gage = 29.7 - 14.7 = **15 psig**

Answer is (D)

MA3:

The sketch shows the conditions of the problem.

$$\frac{P_A}{\rho_A} + Z_A + \frac{U_A^2}{2g_c} - F + w = \frac{P_B}{\rho_B} + Z_B + \frac{U_B^2}{2g_c}$$

$P_A = P_B \quad \rho_A = \rho_B \quad U_A = 0 \quad$ cross sectional area of tank very large

∴ The Bernoulli equation reduces to

$$Z_A + W - F = Z_B + \frac{U_B^2}{2g_c}$$

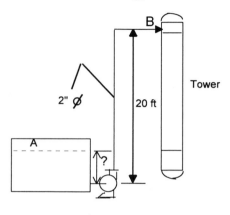

$$Z_A = Z_B + \frac{U_B^2}{2g_c} + F - W$$

$F = 1.5$ ft $\quad Z_B = 20$ ft

$U_B = 80/(7.48 \times 60 \times 0.0233) = 7.65$ ft/s

$$\frac{U_B^2}{2g_c} = 7.65^2/64.4 = 0.91 \text{ ft}$$

$$\frac{m W_P}{550} = 0.25 \qquad m = \frac{80 \times 62.4 \times 1.0}{7.48 \times 60} = 11.12 \text{ lb/s}$$

$W_P = 0.25(550)/11.12 = 12.36$ ft

Therefore, $Z_A = 1.5 - 12.36 + 20 + 0.91 =$ **10.05 say 10 ft**.

The answer is (B)

MA4:

P_A = 20 psia
u_A = 15 ft/s
Z = 0 ft datum

Z = -20 ft
P_B = ?
U_B = 10 ft/s

The Bernoulli equation can be written for the conditions of the problem as

$$\frac{P_A}{\rho} + Z_A + \frac{U_A^2}{2g_c} + q = \frac{P_B}{\rho} + Z_B + \frac{U_B^2}{2g_c}$$

By substituting the given data, the equation becomes

$$\frac{20 \times 144}{56.1} + 0 + \frac{15^2}{64.4} + 0.005 \times 778 = \frac{P_B \times 144}{56.1} - 20 + \frac{10^2}{64.4}$$

Solving for P_B, P_B = 30.06 psia say **30 psia**.

Answer is (B)

Corrosion:

MA5:

From the point of corrosion and cost point, Haveg seems more suitable. The environment in the hydrolyzer is strongly acidic and contains phosphoric acid, sulfuric acid and hydrochloric acid. The pressure of operation is near atmospheric.

Answer is (B)

MA6:

In the hydrolyzer, there is acidic environment. The temperature is below 300 °F. From the point of corrosion and heat transfer, karbate is more suitable. The thermal conductivity of Teflon is poor. Also, its cost is very high. So also of the other alloys mentioned as alternatives. A karbate block exchanger is therefore to be preferred.

Answer is (B)

MA7:

High temperature and corrosive acid gases are present. For such environment usually acid resistant brick lined vessel is good.

Answer is (A)

MA8:

Coupling of two metals far apart in the galvanic series increases the corrosion of more active metal.

Answer is (B)

Flow in pipes and fittings:

MA9:

Correct answer is C since
(1) losses in fittings, contraction etc. are given in terms of velocity constants

(2) L is straight length of pipe and does not include equivalent length of fittings
(3) f is Moody friction factor and therefore excludes the expression B

Answer is (C)

MA10:

Velocity u in 2" pipe = $\dfrac{40}{7.48 \times 60 \times 0.0233} = 3.83$ ft/s

$\dfrac{Du\rho}{\mu} = \dfrac{0.1723 \times 3.83 \times 62.3}{0.982 \times 0.000672} = 6.23 \times 10^4$

$\epsilon/D = 0.00015/0.1723 = 0.00087$

f = 0.023 from Moody chart

Answer is (B)

MA11:

K = 0.78 for 3" schedule 40 pipe. since liquid enters this pipe. To express in terms pipe, use the following relation:

$K_a = K_b \left(\dfrac{d_a}{d_b}\right)^4$ or $K_a = K_b/\beta^4$ where $\beta = \dfrac{d_b}{d_a} = \dfrac{2.067}{3.068} = 0.6737$

K_b for 2" pipe = $K_a \beta^4$ = 0.78 x (0.6737)^4 = 0.16

Answer is (D)

MA12:

Flow through 1½" pipe is 20 gpm

$u = \dfrac{20}{7.48 \times 60 \times 0.01415} = 3.15$ ft/s

Re No. = $\dfrac{Du\rho}{\mu} = \dfrac{0.1342 \times 3.15 \times 62.3}{0.982 \times 0.000672} = 3.99 \times 10^4$

Answer is (C)

Packed and fluidized beds:

MA13:

$u_{mf} = \dfrac{(\phi_s d_p)^2}{150} \dfrac{(\rho_s - \rho_g)\,g}{\mu} \dfrac{\epsilon_m^3}{1-\epsilon_{mf}}$ for $Re_p < 20$

$= \dfrac{(1 \times 0.322 \times 10^{-3})^2}{150} \dfrac{(62.45 - 0.0752)32.2}{0.018(0.000672)} \dfrac{0.4^3}{1-0.4}$

= 0.01224 ft/s

$$\text{Re}_p = \frac{0.322 \times 10^{-3} \times 0.01224 \times 0.0752}{0.018 (0.000672)} = 0.0245$$

Therefore, minimum allowable velocity, u_{mf} = 0.01224 ft/s

Answer is (B)

MA14:

Velocity at the top of the bed must not exceed u_t for the smallest particle in the bed. Therefore calculate u_t for d_p = 0.000164 ft.
First assume Re_p is < 0.4. Then u_t is given by

$$u_t = \frac{g(\rho_s - \rho_g)d_p^2}{18\mu} = \frac{32.2 (62.45 - 0.0752)(0.000164)}{18 (0.018)(0.000672)} = 0.248 \text{ ft/s}$$

$$\text{Re}_p = \frac{0.164 \times 10^{-3} \times 0.248 \times 0.0752}{18 (0.018\, 0.000672)} = 0.253$$

Therefore, assumption about Re No. is justified.

Answer is (C)

MA15:

$$\text{Fr No.} = \frac{u_{mf}^2}{d_p\, g} = \frac{0.01224^2}{0.322 \times 10^{-3}(32.2)} = 0.0145$$

Answer is (B)

MA16:

Static pressure drop $\Delta P = (1 - \epsilon_{mf})(\rho_s - \rho_g)\frac{g}{g_c} \times L$

= (1 - 0.4)(62.45 - 0.0752)(1)(1)

= 37.44 lb/ft² = 0.26 psi

Answer is (A)

Properties of fluids:

MA17:

$$\text{Kinematic viscosity} = \frac{Absolute\ viscosity}{specific\ gravity} = 3.71 \text{ centistokes (given)}$$

Specific gravity = $\frac{54.2}{62.4} = 0.869$

Therefore, viscosity = 0.869(3.71) = 3.223 cP, $\frac{g}{cm.s}$

$$= 3.223(2.42) = 7.8 \ \frac{lb}{ft.h}$$

Answer is (C)

MA18:

Get properties of steam from steam tables. Solution of problem requires interpolation of data.

	Temperature, °F		
	600	640	700
Pressure, psia	Btu/lb	Btu/lb	Btu/lb
400	1306.9		1362.7
450	1302.8	1326.2	

Enthalpy at 630 °F and P = 400 psia
$$= (1362.7 - 1306.9)(30/100) + 1306.9 = 1323.64 \text{ Btu/lb}$$

Enthalpy at 630 °F and P = 450 psia
$$= (1326.2 - 1302.8)(30/40) + 1302.8 = 1320.35 \text{ Btu/lb}$$

By interpolation between pressures, the enthalpy at 630 °F and P = 430 psia
$$= (1323.35 - 1323.64)(30/50) + 1323.64$$
$$= \mathbf{1321.67} \text{ Btu/lb}$$

Answer is (A)

MA19:

$$T_r = \frac{60 + 273.15}{289 + 273.15} = 0.5926 \qquad T_{br} = \frac{80.1 + 273.15}{289 + 273.15} = 0.6284$$

$$k_l = \left[\frac{2.64 \times 10^{-3}}{78.11^{0.5}}\right]\left[\frac{3 + 20(1 - 0.5926)^{0.67}}{3 + 20(1 - 0.6284)^{0.67}}\right] = 0.0003134 \ \frac{cal}{cm.s.K}$$

$$= \mathbf{0.0758} \text{ Btu/(h.ft.°F)}$$

Answer is (B)

MA20:

$$T_{br} = \frac{273.15 + 101.8}{273.15 + 276.2} = \frac{374.95}{549.35} = 0.6825 \quad P_c = 32.9 \text{ atm.}$$

$$\Delta H_{vb} = 1.093(8.31433)(549.35)\left\{\frac{0.6825[\ln(32.9) - 1]}{0.93 - 0.6825}\right\}$$

$$= 34311 \text{ J/gmol}$$

$$= 34311(0.4308) = 14780 \text{ Btu/lbmol}$$

$$= 14780/102.13 = 144.72 \doteq 145 \text{ Btu/lb}$$

Answer is (C)

HEAT TRANSFER

Condensation:

MA21:

Q_s = Heat duty in the sensible heat transfer section = 30000(286.7 - 251.4)
      = 1059000 Btu/h
Q_c = heat transfer by condensation of vapor = 30000(251.4 - 119.1)
      = 3969000 Btu/h
Q_t = total head load, Btu/h = 1059000 + 3969000
      = 5028000 Btu/h = **5.03 MM Btu/h**

Answer is (B)

MA22:

$$\text{Cooling water circulation rate} = \frac{5.03 \times 10^6}{(1)(100 - 65)(500)} = 287.5 \text{ GPM}$$

$$= 143750 \text{ lb/h}$$

$$\text{Rise in temperature of cooling water during condensation} = \frac{3.969 \times 10^6}{(1)143750} = 27 \text{ °F}$$

$$\text{Rise in temperature of cooling water during desuperheating} = \frac{105900}{(1)(143750)} = 27 \text{ °F}$$

Calculate LMTDs

vapor °F		200	126	125
cooling water °F		100	92.6	65
ΔT °F		100	33.4	60

LMTD °F $\text{LMTD} = \dfrac{100-33.4}{\ln \frac{100}{33.4}} = 59.82$ °F $\text{LMTD} = \dfrac{60-33.4}{\ln \frac{60}{33.4}} = 45.4$ °F

$\dfrac{Q_{ds}}{\Delta T_{ds}} = \dfrac{1059000}{59.82} = 17703$ $\dfrac{Q_c}{\Delta T_c} = \dfrac{3969000}{45.4} = 87423$

Weighted temperature difference, $\Delta T = \dfrac{5030000}{17703 + 87423} = 47.9$ °F

Answer is (A)

MA23:

Average temperature of cooling water, $t_a = \dfrac{65+100}{2} = 82.5$ °F

$h_{io} = 840 \times \dfrac{D_i}{D_o} = 840 \times \dfrac{0.62}{0.75} = 694.4$ Btu/h.ft². °F

$\bar{h}_o = 150$ Btu/h.ft².°F (given)

Vapor temperature, $t_{sv} = \dfrac{126 + 125}{2} = 125.5$ °F

Wall temperature, $t_w = 82.5 + \dfrac{150}{694.4 + 150} \times (125.5 - 82.5) = 90.1$ °F

Condensate film temperature = $t_f = t_{sv} - \dfrac{3}{4}(t_{sv} - t_w)$

= 125.5 - 0.75 (125.5 - 90.1)

= 99 °F

Answer is (C)

MA24:

Assume condensate density and thermal conductivity to be constant over the temperature range in this problem because of very small changes.
For initial trial assume $h_o = 150$ Btu/h.ft². °F

$L_c = 10$ ft given. $G'' = \dfrac{W}{L_c N_t^{\frac{2}{3}}} = \dfrac{30000}{10 \times 352^{\frac{2}{3}}} = 59$ lb/(h.lin. ft)

As calculated in MA23 above $t_f = 99\ °F$

At this temperature, $\mu_f = \frac{100-99}{100-90} \times (0.155 - 0.148) + 0.141$

$\qquad\qquad = 0.142$ cP $= 0.3446$ lb/ft.h

$\left(\frac{4G"}{\mu_f}\right)^{-\frac{1}{3}} = \left(\frac{4\times 59}{0.3446}\right)^{-\frac{1}{3}} = 0.113$ $\left(\frac{\mu_f^2}{k_f^3 \rho_f^2 g}\right)^{-\frac{1}{3}} = \left(\frac{0.3446^2}{0.052^3 \times 34.3^2 \times 4.18 \times 10^8}\right) = 848.7$

$\bar{h} = 1.5 \left(\frac{4G"}{\mu_f}\right)^{-\frac{1}{3}} \left(\frac{\mu_f^2}{k_f^3 \rho_f^2 g}\right)^{-\frac{1}{3}}$ $= 1.5(0.13)(848.7) = 165.5$ Btu/h.ft². °F

Recalculate the film temperature.

Wall temperature, $t_w = 82.5 + \frac{165.5}{694.4 + 165.5} \times (125.5 - 82.5) = 90.8$ °F

Condensate film temperature $t_f = t_{sv} - \frac{3}{4}(t_{sv} - t_w)$
$\qquad\qquad = 125.5 - 0.75(125.5 - 90.8)$
$\qquad\qquad = 99.5$ °F vs 99 °F as assumed before

This will not change the viscosity value by much and $\bar{h} = 165$ Btu/h.ft².°F

Answer is (C)

Convection:

MA25:

$t_b = \frac{80+100}{2} = 90$ °F $\mu_b = 18$ cP $D_i = 0.902$ in $= 0.0752$ ft

$N_{Re} = \frac{0.0752 \times 4.5 \times 56}{18 \times 0.000672} = 1567$ This is streamline flow.

$\frac{h_i D_i}{k_b} = 1.86 \left[\left(\frac{D_i G}{\mu_b}\right)\left(\frac{C_{pf} \mu_b}{k_b} \frac{D_i}{L}^{\frac{1}{3}}\right)\right]\left(\frac{\mu_b}{\mu_w}\right)^{0.14}$

$\frac{D_i G}{\mu_w} = 1567$

$\frac{C_P \mu_b}{k_b} = \frac{0.48 \times 18 \times 2.42}{0.08} = 261.4$

$\frac{\mu_b}{\mu_w} = \frac{18}{3.6} = 5$ $\frac{k_b}{D_i} = \frac{0.08}{0.0752} = 1.0638$

$\frac{D_i}{L} = \frac{0.0752}{12} = 0.00627$

$$h_c = 1.86 \times (1.0638)[1567 \times 261.4 \times 0.00627]^{\frac{1}{3}}(5)^{0.14}$$

= **34.1** Btu/h.ft². °F

Answer is (C)

MA26:

No. of tubes per pass = 172/4 = 43 tubes
Cross sectional area = 0.546 in² x 43/144 = 0.163 ft²
Mass velocity through tubes G = 154000/0.163 = 944785 lb/h.ft²
Dittus-Boelter equation uses properties at bulk temperature of the fluid.

$$t_b = \frac{80+120}{2} = 100 \text{ °F} \quad D_i = 0.834/12 = 0.0695 \text{ ft}$$

$\mu = 1.66$ lb/h.ft   k = 0.363 Btu/h.ft².°F $\rho = 62.3$ lb/ft³

$$\left(\frac{D_i G}{\mu}\right)^{0.8} = \left(\frac{0.0695 \times 944785}{1.66}\right)^{0.8} = 4762$$

$$\left(\frac{C_P \mu}{k}\right)^{0.4} = \left(\frac{1 \times 1.66}{0.363}\right)^{0.4} = 1.837 \quad \text{since water is being heated.}$$

Dittus-Boelter equation is

$$\frac{h_i D_i}{k} = 0.023 \left(\frac{D_i G}{\mu}\right)^{0.8} \left(\frac{C_P \mu}{k}\right)^{0.4} \quad \text{for heating of fluid.}$$

Therefore, $h_i = 0.023 \left(\frac{k}{D_i}\right)\left(\frac{D_i G}{\mu}\right)^{0.8}\left(\frac{C_P \mu}{k}\right)^{0.4}$

$$= 0.023 \left(\frac{0.363}{0.0695}\right)(4762)(1.837)$$
$$= \textbf{1050 Btu/h.ft².°F}$$

Answer is (B)

MA27:

Shell side flow area = $a_s = \frac{ID \times C \times B}{P_T \times 144} = \frac{21.25 \times 0.25 \times 6}{1.25 \times 144} = 0.1771$ ft²

$$G_s = \frac{100000}{0.1771} = 564706 \text{ lb/h.ft²}$$

Answer is (A)

MA28:

Shell side heat transfer coefficient is given by

$$\frac{h_o D_e}{k} = 0.36\left(\frac{D_e G_s}{\mu}\right)^{0.55}\left(\frac{C_p \mu}{k}\right)^{\frac{1}{3}}\left(\frac{\mu}{\mu_w}\right)^{014}$$

Equivalent diameter on the shell side, d_e for triagular pitch

$$= \frac{4\left(\frac{1}{2}P_T \times 0.86 P_T - \frac{1}{2}\frac{\pi d_0^2}{4}\right)}{\frac{1}{2}\pi d_o} = \frac{4(0.5 \times 1.25 \times 0.86 \times 1.25 - 0.5 \times \pi \times 1^2/4)}{0.5 \times \pi \times 1}$$

$$= 0.7109 \text{ in} = 0.0592 \text{ ft}$$

$t_B = \frac{190+120}{2} = 155$ °F $\mu = 0.76$ cP $= 0.76(2.42) = 1.84$ lb/h.ft

$$\left(\frac{D_e G_s}{\mu}\right)^{0.55} = \left(\frac{0.592 \times 564706}{1.84}\right)^{0.55} = 220$$

$$\left(\frac{C_p \mu}{k}\right)^{\frac{1}{3}} = \left(\frac{0.88 \times 1.84}{0.34}\right)^{\frac{1}{3}} = 1.68$$

Because of low viscosity value, assume $\mu/\mu_w \simeq 1$

Then, $h_o = 0.36(0.34/0.0592)(220)(1.68) = 769$ Btu/h.ft² .°F

Answer is (C)

Fouling:

MA29:

Assume tube wall thickness is very small and therefore $A_o/A_i \cong 1.0$

Also, assume negligible metal wall resistance. Then,

$$\frac{1}{U_C} = \frac{1}{h_i}\frac{d_o}{d_i} + \frac{1}{h_o} = \frac{1}{700}\frac{1}{0.902} + \frac{1}{250} = 0.005584$$

$U_C = 1/0.005584 = 179$ Btu/h.ft².°F

And $\frac{1}{U_D} = \frac{1}{U_C} + R_{di} + R_{do} = \frac{1}{179} + 0.001 + 0.0015 = 0.008084$

$U_D = 123.7$ Btu/h.ft².°F

Answer is (B)

MA30:

From the Wilson plot, the equations for the lines are

$$\frac{1}{U_0} = \frac{1}{C}\frac{1}{u^{0.8}} + 0.00092 \quad \text{for the dirty tube}$$

and $\quad \frac{1}{U_0} = \frac{1}{C}\frac{1}{u^{0.8}} + 0.0004 \quad$ for the clean tube

The intercept in case of dirty tube consists of three resistances

Intercept (dirty tube) = $R_v + R_w + R_d$

Intercept (clean tube) = $R_v + R_w$

Where R_v = Resistance of steam film,   h.ft².°F/Btu
$\quad\quad\quad R_w$ = Resistance of wall,   h.ft².°F/Btu
$\quad\quad\quad R_d$ = Resistance due to fouling,   h.ft².°F/Btu

At a velocity of 1 ft/s, the reciprocal of the slope of either line gives the value of the inside film coefficient based on the outside area of tube.

Hence, since slope of the lines is 1/C = 0.0038,

$h_{i\,o}$ = 1/0.0038 = 263.2 Btu/h.ft².°F

and h_i based on inside area = 263.2(1/0.902) = **291.8** Btu/h.ft².°F

Answer is (D)

MA31:

Intercept of the line for the clean tube = $R_v + R_w$ = 0.0004

Tube wall resistance = 0.000068

Then R_v = 0.0004 - 0.000068 = 0.000332
Steam film coefficient = 1/0.000332 = **3012** Btu/h.ft².°F

Answer is (A)

MA32:

Intercept of the line for the dirty tube = $R_v + R_w + R_d$

$\quad\quad\quad\quad\quad\quad\quad\quad\quad\quad\quad\quad\quad\quad\quad$ = 0.00092 from the Wilson plot

Intercept of the line for the clean tube = $R_v + R_w$ = 0.0004

Therefore, R_{do} = 0.00092 - 0.0004 = 0.00052

Then h_{do} = 1/0.00052 = 1923 Btu/h.ft².°F based on outside area

h_{di} = 1923(1/0.902) = **2132** Btu/h.ft².°F

Answer is (D)

Insulation:

MA33:

Resistance of the two insulation layers

$$= \frac{1.25/12}{0.058\left(\pi \frac{3.48}{12}\right)} + \frac{2.5/12}{0.042\left(\pi \frac{7.07}{12}\right)}$$

$$= 1.9713 + 2.68 = \mathbf{4.651}\ \text{h.ft}^2.\text{°F/Btu}$$

Answer is (A)

MA34:

Resistance of pipe wall = $R_w = \frac{0.154/12}{26\left(\pi \frac{2.22}{12}\right)} = 0.00085$ h.ft².°F/Btu

Heat loss per foot of pipe = $q = \frac{900 - 120}{0.00085 + 1.971 + 2.68} = 167.7$ Btu/h.ft of pipe.

Answer is (C)

MA35:

Temperature drop is directly proportional to the resistance.

Resistance upto interface of insulation layers = 0.00085 + 1.971 = 1.97185
Total resistance upto surface of second layer = 0.00085 + 1.971 + 2.68 = 4.65185
Temperature drop upto interface of insulation layers

$$= \frac{1.97185}{4.65185} \times (900 - 120) = 330.6\ \text{°F} \doteq 331\ \text{°F}$$

Interface temperature between the two insulation layers = 900 - 331 = **569** °F

Answer is (A)

MA36:

Surface coefficient of heat transfer, h_a

$$= \frac{q}{A_o \Delta t} = \frac{167.7}{\pi \left(\frac{9.87}{12}\right)(1)(120-85)} = \mathbf{1.85} \quad \text{Btu/h.ft}^2.°F$$

$h_a = h_c + h_r$

$h_c = 0.5 \left(\frac{\Delta t}{d_o}\right)^{0.25} = 0.5 \left(\frac{120-85}{2.375}\right)^{0.25} = 0.98$ Btu/h.ft².°F

Therefore $h_r = 1.85 - 0.98 = \mathbf{0.87}$ Btu/h.ft².°F

Answer is (A)

PLANT DESIGN

Economics:

MA37:

If i is the rate of return, the future worth of all the end-of-year incomes is

$$35000(1+i)^4 + 37000(1+i)^3 + 43000(1+i)^2 + 48000(1+i) + 52000$$

This also equals $(120000 + 15000)(1+i)^5 - 15000 - 12000$

By equating the two, and dividing by 1000, one gets

$$35(1+i)^4 + 37(1+i)^3 + 43(1+i)^2 + 48(1+i) + 52 = (135)(1+i)^5 - 27$$

or $35(1+i)^4 + 37(1+i)^3 + 43(1+i)^2 + 48(1+i) + 79 - (135)(1+i)^5 = 0$

Assume i = 15 %   LHS = 30.972
         i = 20 %   LHS = - 3.288

The correct rate of return is closer to 20 %,
One more trial with i = 19.9 brings LHS closer to 0.

Answer is (B)

MA38:

Assume i = 20 %

Using appropriate factors from tables of data provided, we can write

Present worth

$$= 35000(0.9063) + 37000(0.7421) + 43000(0.6075) + \\ 48000(0.4974) + 52000(0.4072) + 27000(0.3679)$$
$$= \$140283 \quad \text{Need } (\$12000 + 15000) = \$135000$$

Assume $i = 25\%$,

Present worth

$$= 35000(0.8848) + 37000(0.6891) + 43000(0.5367) + \\ 48000(0.4179) + 52000(0.3255) + 27000(0.2865)$$

$$= \$124264$$

Find approximate return rate by interpolation.

% return = $\frac{135000 - 140283}{124264 - 140283} \times 5 + 20 = 21.65\%$

Answer is closer to 21.7 % (for exact figure some more trials will be required)

Answer is (D)

MA39:

For finding out the optimum thickness, it is necessary to plot the total cost against the insulation thickness. The minimum cost point on the total cost curve will give the optimum thickness. to get total costs, read the fixed cost and cost due to heat loss from the curves and by addition get the total cost at various thicknesses as follows

insulation thickness, in	1	2	3	4	5
Annual fixed cost $	215	355	485	603	735
Cost due to heat loss $	910	610	510	460	425
Total annual costs $	1125	965	995	1063	1160

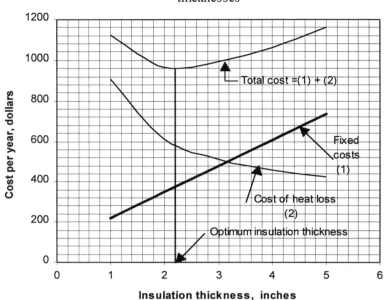

Fixed costs and costs of heat loss at various thicknesses

These values are plotted in the figure provided with the problem. On the total cost curve, the minimum cost occurs at insulation thickness of 2.25 inches approximately

Answer is (A)

MA40:

Profit P = Cost of product realized - cost of reactant converted - operating cost

In case of parallel reactions, products formed are in ratio of their rate constants.

Molls of A converted = $F_{AA} A_X$ unreacted A is recovered and recycled.

Molls of R formed = $\dfrac{k_1}{k_1+ k_2}(F_{A0}X_A) = \dfrac{4}{4+1}(F_{A0}X_A) = 0.8\,(F_{A0}X_A)$

Profit P = $(0.8\ F_{AA}\ A_X)(5\ \$) - F_{AX}(1.5\ \$) - 1.5\ F_{AA} - 30$

Eliminate F_{AA} from the equation by replacing the value of F_{AA} in terms of A_X

$$F_{AA} = \dfrac{V k\, C_{Ao}(1 - X_A)}{X_A} = \dfrac{60\,(4+1)(1)\,(1 - X_A)}{X_A} = 300\,\dfrac{1- X_A}{X_A}$$

Substituting in the equation for net profit,

$$P = 4.0\left[\frac{300(1-X_A)}{X_A}\right]X_A - 1.5\left[\frac{300(1-X_A)}{X_A}\right]X_A - 1.5\frac{300(1-X_A)}{X_A} - 30$$

which reduces on simplification to

$$P = 1200(1-X_A) - 450(1-X_A) - 450\frac{(1-X_A)}{X_A} - 30$$

For maximum profit, differentiate with respect to A_X,

$$\frac{dP}{dX_A} = 1200(-1) - 450(-1) - 450\left[\frac{X_A(-1) - (1-X_A)(1)}{X_A^2}\right] - 0 = 0$$

or $\quad -750 X_A^2 = -450 \quad$ or $\quad A_X = 0.7746 \cong 77.5$

Answer is (B)

END OF SOLUTIONS TO PM EXAM 1..

SOLUTIONS TO PM SAMPLE EXAM 2
FLUIDS
Control of flow systems:

AA1:

$$C_{vc} = 1.16 \times 12.24 \sqrt{\frac{0.788}{0.75/1.02}} \cong 14.7 \text{ at max. flow}$$

$$C_{vc} = 1.16 \times 4.2 \sqrt{\frac{0.788}{2.03/1.02}} \cong 3.1 \text{ at min. flow}$$

Required rangeability = $\frac{C_{vc} \text{ at maximum flow}}{C_{vc} \text{ at minimum flow}} = \frac{14.7}{3.1} = 4.74$

This is closest to 5:1.

Answer is (B)

AA2:

To fill the missing numbers for line and equipment losses, assume the pump curve is flat. Therefore the discharge pressure applicable to second branch is 7.89 kg/cm². Then allowable control valve pressure drop [kg/cm²] in Branch #2 is:

CV pressure drop = 7.89 - 3.5 - 0.42 - 1.68 = 2.29 kg/cm²

Answer is (C)

AA3:

Numbers for pressure drops due to friction and equipment are missing. First calculate these losses. For this, use the fact that pressure drop is proportional to square of velocity or flow rate.

Frictional loss = $\left(\frac{17.4}{14.5}\right)^2 \times 0.42 = 0.61$ kg/cm²

Equipment loss = $\left(\frac{17.4}{14.5}\right)^2 \times 1.68 = 2.42$ kg/cm²

Therefore CV pressure drop at maximum flow in Branch #2 is:

CV pressure drop = 7.89 - 3.5 - 0.61 - 2.42 = 1.36 kg/cm²

Answer is (D)

AA4:

$$C_{vc} = 1.16(17.4)\sqrt{\frac{0.788}{1.36/1.02}} = 15.5$$

If the valve size 1½" is selected, $C_v = 21$

Then C_v/C_{vc} = **21/15.5 = 1.35**

This lies between 1.25 to 2 which is required for good operation of the control valve.

Answer is (B)

Economics:

AA5:

Present value, P = $ 100,000
Salvage value, L = $ 10,000
Useful life, n = 10 years

Straight line depreciation = $\frac{1000 - 10000}{10}$ = **$9000** per year.

Book value after 5 years = 100000 - 9000(5) = **$55,000**

Answer is (A)

AA6:

For declining balance method,

$$f = 1 - \left(\frac{L}{P}\right)^{\frac{1}{n}} = 1 - \left(\frac{10000}{100000}\right)^{\frac{1}{10}} = 0.2057$$

Book value after 5 years, $B = P(1-f)^a = 100000(1-0.2057)^5$

= **$ 68383**

Answer is (B)

AA7:

In double decline balance method, generally no salvage value is allowed.

Therefore, depreciation rate = 2 × (*rate of straight line depreciation*)

$$= 2(0.1) = 0.2$$
Book value $B = 100000(1 - 0.2)^5 = \$32768$

Answer is (C)

AA8:

By sum of digits,

$$B = P - \frac{(P-L)r}{n(n+1)} \times [2n - (r-1)]$$

$$= 100000 - \frac{90000(5)}{10 \times 11}[2 \times 10 - (5-1)]$$

$$= \$ 34545$$

Answer is (C)

Sensors:

AA9:

Velocity through the orifice $u_o = \frac{34.1}{3600 \times 0.001553} = 6.1$ m/s

$$\frac{Du\rho}{\mu} = \frac{0.0445(6.1)(1000)}{1 \times 10^{-3}} = 2.72 \times 10^5$$

Therefore, flow is fully turbulent.

Hence $C_o = 0.61$ $\beta = 1.75/3.068 = 0.5704$ $1 - \beta^4 = 0.894$

Hence, $6.1 = 0.61\sqrt{\frac{2(9.81)(\Delta H)}{0.804}}$ Then $\Delta H = 4.56$ $m = 456$ cms

$$\Delta H = h_m(\frac{13.6}{1} - 1)$$

$h_m = \Delta H/12.6 = 456/12.6 = 36.2$ cms

Answer is (C)

AA10:

Span of instrument = 60 - 0 = 60 kg/cm²

Pneumatic signal is from 20 to kPa

Therefore the measured pressure = $\frac{70-20}{100-20} \times 60 = 37.5$ kg/cm²

Answer is (C)

AA11:

When signal is 15 mA dc, $\Delta H = \frac{15-4}{20-4} \times 220 = 151.3$ cms

When $\Delta H = 220$ cms Flow rate = 15 m³/h

Now flow = $A_O u_O = A_O C_O \sqrt{\frac{2g\Delta H}{1-\beta^4}} = K\sqrt{\Delta H}$

$\frac{Flow\ 2}{Flow\ 1} = \sqrt{\frac{\Delta H_2}{\Delta H_1}} = \sqrt{\frac{151.3}{220}} = 0.83$

Flow2 = 0.83(Flow1) = 0.83(15) = 12.45 → 12.5

Answer is (B)

AA12:

$\Delta t\ (initial) = 100^0 C$

Δt after 6 seconds, $\Delta t = 100(1-e^{-t/\tau}) = 100(1-e^{-6/2}) = 95$ °C
t = 900 + 95 = 995 °C

Answer is (B)

Pumps and turbines:
AA13:

Suction static head = 25.5 x 0.433 x 0.68 = **7.5** psi

Answer is (A)

AA14:

Origin pressure	psia	=	110.0
Suction static head	psi	=	7.5
- Line loss	psi	=	1.5
Pump suction pressure	psia	=	116.0
- Vapor pressure	psia	=	110.0
NPSH available	psi	=	6.0
NPSH available 6 x 2.31/0.68 = 20.4 ft		=	20.4

Answer is (A)

AA15:

$$\text{Discharge pressure} = 180 + 55.4 + 7.3 + 10.4 + 12.1 + 4.8 + 2.4 = 272.4 \text{ psia}$$

Total dynamic head = discharge pressure - suction pressure
$$= 272.4 - 116$$
$$= 156.4 \text{ psi} = 156.4 \times 2.31/0.68 = 531.3 \text{ ft}$$

Answer is (D)

AA16:

$$\text{BHP} = \frac{GPM \times psi}{1714 \times 0.65} = 203.8$$

Answer is (A)

HEAT TRANSFER

Resistance:

AA17:

$$\frac{q}{A} = \frac{k_m \Delta t}{\Delta X} = \frac{0.95(1800 - 1720)}{6/12} = 152 \text{ Btu/h.ft}^2$$

$$Q = 152(450)(24) = 1.6416 \times 10^6 \text{ Btu/day}$$

Answer is (D)

AA18:

Assume steady state. Therefore 152 Btu/h.ft² pass through each section

For insulating brick,

$$\Delta X = \frac{0.14(1720 - 280)}{152} = 1.3263 \text{ ft} = 15.92'' \simeq 16''$$

Answer is (A)

AA19:

$$R_T = \frac{1}{h_i}\frac{D_O}{D_i} + r_i\frac{D_O}{D_i} + \frac{l_w}{k_w}\frac{D_O}{D_{av}} + r_o + \frac{1}{h_O}$$

$$= \frac{1}{2010}\frac{1}{0.902} + 0.001\frac{1}{0.902} + \frac{0.049/12}{63}\frac{1}{0.951} + 0.0005 + \frac{1}{1840}$$

$$= 0.0005516 + 0.001109 + 0.00006815 + 0.0005 + 0.0005435$$

$$= 0.00277225$$

$$\frac{1}{U_O} = 0.0027725 \qquad U_O = 360.7 \text{ Btu/(h.ft}^2.°\text{F)}$$

Answer is (A)

AA20:

Because of cleaning the tubes on both sides, the fouling resistances are eliminated. Therefore

$$R_T = \frac{1}{h_i}\frac{D_o}{D_i} + \frac{l_w}{k_w}\frac{D_O}{D_{av}} + \frac{1}{h_O}$$

$$= \frac{1}{2010}\frac{1}{0.902} + \frac{0.049/12}{63}\frac{1}{0.951} + \frac{1}{1840}$$

$$= 0.0005516 + 0.00006815 + 0.0005435$$

$$= 0.00116325$$

$$\frac{1}{U_o} = 0.00116325 \qquad U_O = 859.7 \text{ Btu/(h/ft}^2.°\text{F)}$$

Answer is (B)

Conduction:

AA21:

Thickness of pipe wall = (2.375 - 2.067)/2 = 0.154 in

Since thickness is very is very small, arithmetic mean can be used.

Average diameter = (2.375+ 2.067)/2 = 2.221 in.

Resistance of steel wall = $\frac{0.154/12}{26\,(2.221\pi/12)} = 0.00085$ (h.ft^2.°F)/Btu

Answer is (A)

AA22:

Logarithmic average of diameters need be used.
Inner diameter of magnesia covering = 2.375 in.
Outer diameter of magnesia covering = 2.375 + 3 = 5.375 in
Outer diameter of cork = 5.375 + 5 = 10.375 in

Average diameter for magnesia layer = $\frac{5.375 - 2.375}{\ln \frac{5.375}{2.375}} = 3.673$ in

Average diameter for cork layer = $\frac{10.375 - 5.375}{\ln \frac{10.375}{5.375}} = 7.603$ in

$$R_T = \frac{0.154/12}{26(2.221\pi/12)} + \frac{1.5/12}{0.034(3.673\pi/12)} + \frac{2.5/12}{0.03(7.603\pi/12)}$$

$$= 0.00085 + 3.82333 + 3.48886$$

$$= 7.313 \text{ h.ft}^2.{}^\circ\text{F/Btu}$$

Heat loss = q = (900 - 120)/7.313 = 106.7 Btu/h.ft length of pipe

Answer is (B)

AA23:

As calculated in problem HC1-2, total resistance is 7.313 h.ft^2.$^\circ$F/Btu

Answer is (B)

AA24:

Since temperature drop is directly proportional to resistance, temperature drop upto inner surface is based on sum of resistances of pipe wall and magnesia layer.

Resistance of pipe wall and magnesia layer = 0.00085 + 3.82333 = 3.82418

Temp. drop corresponding to this resistance = (3.82418/7.313)(900 - 120) = 408 $^\circ$F

Then temperature at the interface of magnesia and cork = 900 - 408 = 492 $^\circ$F

Answer is (C)

Energy conservation:

AA25:

Heat savings = mC$_P$ Δt = 4250(1)(212 – 120) = 391000 Btu/h

Heat of vaporization = 928.8 Btu/lb from steam tables.

Steam saved = 391000/928.8 = 421 lb/h

Yearly savings = $\dfrac{421\ (8000)\ (5)}{1000}$ = \$16,840 per year

Answer is (A)

AA26:

Temperature of process fluid leaving the exchanger

$$mC_P\Delta t = 391000 \qquad \Delta t = \frac{391000}{5000(1)} = 78.2\ ^0F$$

$t_0 = 100 + 78.2 = 178.2\ ^0F$

$\Delta t_2 = 120 - 100 = 20\ ^0F \qquad \Delta t_1 = 212 - 178.2 = 33.8\ ^0F$

$$\Delta t_{lm} = \frac{33.8 - 20}{\ln\frac{33.8}{20}} = 26.3\ ^0F$$

Area of new exchanger = $\frac{391000}{26.3 \times 100} = 148.7$ ft²

Cost of new exchanger = $(\frac{148.7}{50})^{0.6}(10,000) = \$19,231$

Payback period = 19231/16840 = 1.142 yr = 13.7 months

Answer is (A)

AA27:

Energy available from flue gas = 178500(0.3)(1800 - 500) = 69.62 MM Btu/h

Allow 1 % heat loss. Feed water is at 220 °F.

Heat required to vaporize 1 LB steam = $H_{422} - h_{220}$ = 1203.7 - 188.14 = 1015.56 Btu/lb

Steam produced = $69.62 \times 10^6 \times 0.99/1015.56 = 67,868$ lb/h

Steam cost to company was $2.5 to the company.

Therefore, savings per year = 2.5(67868) (6000)/1000 = $1018020

Answer is (A)

AA28:

Calculate heat loss with additional insulation.

$d_i = 6.625" = 0.552$ ft $\qquad d_o = 6.625 + 6" = 12.625" = 1.052'$

Assume $t_s = 100\ ^0F \quad h_c = 0.5(\frac{30}{12.625})^{0.25} = 0.62$ Btu/h.ft².°F

$$h_r = 0.68(0.174)\left[\left(\tfrac{560}{100}\right)^4 - \left(\tfrac{530}{100}\right)^4\right]/(100-70)$$

$$= 0.77 \text{ Btu/h.ft}^2.{}^\circ F$$

$h_c + h_r = 0.77 + 0.62 = 1.39$ Btu/h.ft². °F

$$q = \frac{377 - 70}{\ln\tfrac{1.052}{0.552}/(2\pi \times 0.05) + 1/(1.39 \times \pi \times 1.052)} = 135 \text{ Btu/h.ft of pipe}$$

Check assumed surface temperature.

$$q_s = 1.39(t_s - 70)(\pi \times 1.052) = 135$$

$t_s - 70 = 29.4 \quad t_s = 70 + 29.4 = 99.4$ vs assumed 100 °F

Heat loss /yr = 135(1000)(8760) = 1182.6 MM Btu/yr
Additional savings based on 75 % boiler efficiency and $2/MM Btu fuel cost

$$= \frac{(2455 - 1182.6)}{0.75} \times 2 = \$3393/yr$$

For a payout period of two years, maximum insulation cost = $\dfrac{2 \times 3393}{1000} = \$6.79/ft$

Answer is (C)

Radiation:

AA29:

$$A_1 = \pi\left(\tfrac{2.38}{12}\right)(1) = 0.622 \text{ ft}^2$$

$A_2 = 4(1)(1) = 4$ ft²

In this case, $F_A = 1$, $F_e = \dfrac{1}{\tfrac{1}{\epsilon_1} + \tfrac{A_1}{A_2}\left(\tfrac{1}{\epsilon_1}-1\right) - 1} = \dfrac{1}{\tfrac{1}{0.8} + \tfrac{0.622}{4}\left(\tfrac{1}{0.28}-1\right)} = 0.606$

$$Q = F_A F_e A(\sigma)\left(T_p^4 - T_d^4\right) = (1)(0.622)(0.173 \times 10^{-8})(760^4 - 530^4)$$

$$= 166 \text{ Btu/h.lin.ft of pipe}$$

Answer is (C)

AA30:

The surroundings are very large, Therefore, radiation reflected back to pipe can be neglected. Also, assume negligible resistance from steam film and metal.

Surface temperature, $t_s = 325 + 460 = 785$ R
Air temperature, $t_s = 70 + 460 = 530$ R

Heat loss from bare pipe = $0.79(0.173 \times 10^{-8})(785^4 - 530^4) = 411$ Btu/h.ft²
Heat loss from painted pipe = $0.35(0.173 \times 10^{-8})(785^4 - 530^4) = 182.1$ Btu/h.ft²

% decrase in radiation loss = $\frac{411-182}{411} \times 100 = 55.7\%$

Answer is (C)

AA31:

Surface area of nickel tube per foot length, $A_1 = \pi(3/12)(1) = 0.785$ ft²
Area of chamber, $A_2 = 4(10/12)(10/12) = 2.778$ ft²
Absorptivity nickel from the radiation source at 1850 °F = its emissivity at 1850 °F = 0.59
Since enclosure is not very large compared to the tube, it is necessary to allow for the emissivity of the silica brick by using the relation

$$\frac{1}{A_1 F_{1\to 2}} = \frac{1}{A_1}\left(\frac{1}{\epsilon_1} - 1\right) + \frac{1}{A_2}\left(\frac{1}{\epsilon_2} - 1\right) + \frac{1}{A_1 \bar{F}_{1\to 2}}$$

Since all the radiation from A_1 is received by A_2, $\bar{F}_{1\to 2} = 1.0$

$$\frac{1}{F_1} = \left(\frac{1}{\epsilon_1} - 1\right) + \frac{A_1}{A_2}\left(\frac{1}{\epsilon_2} - 1\right) + 1 = \left(\frac{1}{0.59} - 1\right) + \frac{0.785}{2.778}\left(\frac{1}{0.78} - 1\right) + 1 = 1.775$$

$F_1 = 1/1.775 = 0.5634$

$q = (0.5634)(0.785)(0.173)(13.1^4 - 23.1^4) = -24,833$ ft²/ft of nickel tube.

Answer is (B)

AA32:

Total emissive power = $E = \int_0^\infty I_\lambda d\lambda$

From this relation, it is apperent that the area under the curve gives the total emissive power of the hot body. Radiation decreases with increase in λ and becomes negligible beyond $\lambda = 16$ microns. ignore the very small radiation beyond $\lambda = 16$ microns and also $\lambda < 0.8$. To get the area under the curve, either use Simpson's rule or count squares. By counting, the area under the curve is 90 x 1 x 0.5 = **45** Btu/h.ft².

Answer is (B)

MASS TRANSFER

Flooding and pressure drop:

AA33:

$$l_w = 0.7267(5) = 3.635 \text{ ft} = 43.6''$$

$$\frac{\text{plate thickness}}{\text{hole diameter}} = \frac{0.0825}{0.25} = 0.33$$

Weir crest height = $h_{ow} = 0.48(F_w)\left(\frac{Q_w}{l_w}\right)^{0.67}$

[Where F_w is correction for effective weir length to be obtained from figure 18-16, Perry's Handbook 6th ed. p 18-11 as follows]

$$\frac{q}{(L_w)^{2.5}} = \frac{120.2}{3.635^{2.5}} = 4.78$$

From figure 18-16 abovementioned, $F_w = 1.04$

$$h_{ow} = 0.48(1.04)\left(\frac{120.2}{43.6}\right)^{0.67} = 0.99 \doteq 1 \text{ in}$$

$$F_{ga} = 3.8\sqrt{0.256} = 1.92$$

From figure 18-15, Perry's Hand Book p 18-10, aeration factor $\beta = 0.59$
Pressure drop due to liquid depth on tray and crest = $0.59(2 + 1) = $ **1.77 in**

Answer is (A)

AA34:

Hole pressure drop is given by $h_o = 0.186 \frac{\rho_g}{\rho_L}\left(\frac{U_h}{C_o}\right)^2$

For $A_h/A = 0.1$, $C_o = 0.73$ from figure 18-14 on p 18-9 Perry's Hand Book 6th ed.

$$U_h = \frac{60}{19.74(0.1)} = 30.4 \text{ ft/s}$$

$$h_o = 0.186 \frac{0.256}{47.8}\left(\frac{30.4}{0.73}\right)^2 = 1.73''$$

Answer is (C)

AA35:

Head loss required to overcome surface tension effect

$$= \frac{0.04\sigma}{\rho_L d_h} = \frac{0.04(18)}{47.8 \times 0.25} = 0.06 \text{ in}$$

Answer is (D)

THERMODYNAMICS

AA36:

Down comer head loss:

Clearance under down comer = 2 inches

Minimum area of flow = $\frac{2 \times 43.6}{144} = 0.606 \text{ft}^2$

$h_d = 0.03(\frac{120.2}{100 \times 0.606})^2 = 0.12"$

Answer is (B)

AA-37:

m = number of payments in a year.

j = monthly interest rate = 0.065/12 = 0.005417

Effective annual interest rate = $\left(1 + \frac{j}{m}\right)^n - 1 = (1 + \frac{0.005417}{12})^{12} - 1 = 0.066972$

$= 6.6972\%$

Answer is (D)

AA-38:

Monthly interest rate = 0.006972

We need amount of monthly payment

n = 30(12) = 360   number of interest periods.

Thus,   200000 = A(P/A, i = 0.005417, 360)

(P/A, i = 0.005417, 360) = $\frac{(1.005417)^{360} - 1}{0.005417 \times (1.005417)^{360}} = 158.2042^*$

* Can be obtained from tables also.

Or   200000 = A(158.2042)

Therefore A = $\frac{200000}{158.2042} = \$1264.19$

Answer is (A)

AA-39:

Straight line depreciation = 10(3350) = $33500

$$A = F(A/P, i, n) = F\frac{i}{(1+i)^n - 1} = 33500\left[\frac{0.065}{(1.065)^{10} - 1}\right]$$

$$= \$2482.51$$

Answer is (D)

AA-40:

$$\text{Capitalized cost} = P + \frac{A}{i} + \frac{(P-L)(A/F, i, n)}{i}$$

Applying to present case,

$$\text{Capitalized cost} = P + P\frac{(A/F, i, n)}{i}$$

$$= P + \frac{P}{i}\left[\frac{i}{(1+i)^n - 1}\right]$$

$$= P\left[1 + \frac{1}{(1+i)^n - 1}\right]$$

$$100000 = P\left[1 + \frac{1}{(1+0.10)^{10} - 1}\right] = P(1.627454)$$

$$P = 100000/1.627454 = \$61445.67$$

Answer is (C)

END OF SOLUTIONS OF PM EXAM 2..

SOLUTIONS TO PM SAMPLE EXAM 3
FLUIDS

Economics:

MB1:

Annual profit before income taxes = 960000 - 620000 = 340,000

Percent return on total initial investment before income taxes

$$= \frac{340000}{1100000 + 120000} \times 100 = 27.9\,\%$$

Answer is (B)

MB2:

Annual profit after income taxes = (340000)(0.52) = 176800

Annual percent return on total initial investment after income taxes

$$= \frac{176800}{1100000 + 120000} \times 100 = 14.49\,\%$$

Answer is (D)

MB3:

Minimum profits required per year before income taxes

$$= (1100000 + 120000)(0.15) = \$183000$$

Fictitious expenses based on capital recovery with minimum profit

$$= 620000 + 183000 = \$803000$$

Annual % return on the total investment based on capital recovery with minimum 15 % return before income taxes

$$\frac{960000 - 803000}{110000 + 120000} \times 100 = 12.87\,\%$$

Answer is (D)

MB4:

Average investment assuming straight-line depreciation and zero salvage value

$$= \frac{1100000}{2} + 120000 = \$670000$$

Annual % return on average investment before income taxes

$$= \frac{340000}{670000} \times 100 = 50.75\ \%$$

Answer is (B)

KINETICS

Biochemical reactors:

MB5:

Michaelis-Menton equation can be written by rearrangement as

$$\frac{1}{-r_A} = \frac{C_A + M}{k_3 C_A C_{E0}} = \frac{M}{k_3 C_{E0}} \frac{1}{C_A} + \frac{1}{k_3 C_{E0}}$$

A plot of $-\frac{1}{r_A}$ vs $\frac{1}{C_A}$ should be a straight line if Michaelis-Menton is to apply. The plot given is a straight line. Slope of the line is $\frac{M}{k_3 C_{E0}}$ and its intercept is $\frac{1}{k_3 C_{E0}}$

Intercept of the line = $\frac{1}{k_3 C_{E0}}$ $k_3 = \frac{1}{4.7728 \times 0.01} = 20.95\ h^{-1}$

Answer (A)

MB6:

Slope of the line = $\frac{M}{k_3 C_{E0}}$ = 1.1932

Therefore, M = 1.1932($k_3 C_{E0}$) = 1.1932(20.95)(0.01) = **0.25**

Answer (C)

MB7:

The Michaelis-Menton rate expression $\frac{ds}{dt} = \frac{-r_{max}\ s}{K_M + s}$

This can be integrated with s(0) = s_0 to give $r_{max} t = s_0 - s + K_m \ln(s_0/s)$

To calculate r_{max}, substitute given data in this equation.

When t = 2 min, conversion = 5%

Then s = 3 x10^{-5} x (1 - 0.05) = 2.85x10^{-5} M at 2 min.

Therefore, $r_{max}.(2) = 3 \times 10^{-5} - 2.85 \times 10^{-5} + 1 \times 10^{-3} \ln(3 \times 10^{-5}/2.85 \times 10^{-5})$

$$= 0.15 \times 10^{-5} + 10^{-3} \times 0.0513$$
$$= (0.15 + 5.13) \times 10^{-5} = 5.28 \times 10^{-5}$$

$r_{max} = 5.28 \times 10^{-5}/2 = 2.64 \times 10^{-5}$

Answer is (C)

MB8:

Using r_{max} from found in KA2-3 above, we write as follows
$2.64 \times 10^{-5}(t) = s_0 - s + K_m \ln(s_0/s)$

$s_0 = 3 \times 10^{-5}$ M and $K_m = 1 \times 10^{-3} = 100 \times 10^{-5}$ M

Therefore, $2.64 \times 10^{-5} (30) = 3 \times 10^{-5} - s + 10^{-5}[100 \ln(3 \times 10^{-5}/s)]$

Simplification yields $1117.6 \times 10^{-5} = -s - 10^{-5}[100 \ln(s)]$

A trial and error solution for s is required.

Assume $s = 1.7 \times 10^{-5}$ Right hand side = 1096.3×10^{-5}

Assume $s = 1.5 \times 10^{-5}$ Right hand side = 1109.3×10^{-5}

Assume $s = 1.4 \times 10^{-5}$ Right hand side = 1116.3×10^{-5}

Assume $s = 1.38 \times 10^{-5}$ Right hand side = 1117.7×10^{-5} close to LHS

s converted = $3 \times 10^{-5} - 1.38 \times 10^{-5} = 1.62 \times 10^{-5}$

Conversion = $(1.62 \times 10^{-5}/3 \times 10^{-5}) \times 100 = $ **54 %**

Answer is (A)

Equilibrium:chemical/phase:

MB9:

$$\Delta G^o_{298} = \Sigma(\Delta \dot{G}^o_{298})_{products} - \Sigma(\Delta G^o_{298})_{reac\tan \tan ts}$$

$$= 2(16.28) + 1(0) - 1(-3.75)$$

$$= \textbf{36.31 Kcal.gmol}$$

Answer is (A)

MB10:

Heat of reaction at 298 K = 2(12.5) + 1(0) - (-29.81) = 54.81 kcal/gmol
Net

$$\Delta C_p^o = 2(0.0028 + 3.00 \times 10^{-5}T) + (0.0069 + 0.4 \times 10^{-5}T) - (0.01178 + 4.268 \times 10^{-5}T)$$

$$= 0.00072 + 2.132 \times 10^{-5}T$$

$$I_H = \Delta H_{298}^o - \Delta a\, T - \tfrac{1}{2}\Delta\beta T^2$$

$$= 54.81 - 0.00072(298) - 1.066 \times 10^{-5}(298)^2 = 53.65$$

T = 500 + 273 = 773 K

$$\Delta H_{773}^o = 53.65 + 0.00072(773) + 1.066 \times 10^{-5}(773)^2$$

$$= \mathbf{60.57}\ \text{Kcal/gmol}$$

Answer is (B)

MB11:

Calculation of K

T = 273 + 704 = 977 K

$$\Delta G_T^o = I_H + I_G T - \Delta a T \ln T - \tfrac{1}{2}\Delta\beta T^2$$

$$36.31 = 53.65 + I_G\, 298 - 0.00072 \ln 298 - 1.066 \times 10^{-5}(298^2)$$

From which $I_G = (36.31 - 53.65)/298 = -0.0509$

$$\Delta G_{977}^o = 53.65 - 0.0509(977) - 0.00072(977)\ln(977) - 1.066 \times 10^{-5}(977^2)$$

$$= -11.097\ \text{kcal/gmol}$$

$$\Delta G_{977}^0 = -RT \ln K_{977}$$

$$\ln K_{977} = -\frac{-11.097 \times 10^3}{1.987 \times 977} = 5.716$$

$$K_{977} = e^{5.716} = \mathbf{303.7}$$

Answer is (C)

MB12:

If X is the conversion, n-butane remaining = 1 - X
C_2H_4 produced = 2X
H_2 produced = X
Total = 1 + 2X

In this case, $K = \dfrac{\left(\dfrac{2X}{1+2X}\right)P_T \left(\dfrac{X}{1+2X}\right)P_T}{\dfrac{1-X}{1+2X}P_T}$ $P_T = 2$ atm.

$$303.7 = \dfrac{4X^2}{(1-X)(1+2X)}$$

Simplification gives $611.4X^2 - 303.7X - 303.7 = 0$

or $X^2 - 0.4967X - 0.4967 = 0$

Solving the equation for X,

$$X = \dfrac{0.4967 \pm \sqrt{0.4967^2 - 4(-0.4967)}}{2} = 0.9956$$

% conversion = 0.9956(100) = **99.56 %**

Answer is (A)

Material selection for reactors:

MB13:

Corrosion rates of carbon steel, type SS 430 and aluminum by phosphoric acid are very high. SS 430 does not have MO as ingredient, which is needed to improve corrosion resistance. SS 317L contains 3 to 4 % Mo and low carbon and has excellent resistance against pitting. It performs very good under oxidizing conditions because of formation of a passive oxide film on the surface. Therefore, it has found application in phosphoric acid production. Select 317L.

Answer is (B)

MB14:

SS 304 and SS 316 both are subject to chloride attack. Reducing conditions and chloride ions the passive film of oxide on the surface. Chloride ions cause pitting and crevice corrosion. When combined with high tensile stresses, stress corrosion can result. Therefore both SS 304 and SS-316 are ruled out.

Alloy 20 has good resistance to sulfuric acid better than stainless steels but is subject to pitting in presence of chloride ions. Therefore, Alloy 20 should be ruled out.

Glass linings are resistant to all concentrations of HCl upto 120 °C. Presence of chloride ions in glass-lined vessels acceptable. Hence choose glassed lined vessel.

Answer is (D)

MB15:

Glass, tantalum are readily attacked by hydrofluoric acid. Therefore, the presence of fluorides rules out selection of these two materials. SS 317L offers good corrosion resistance against wet phosphoric acid, but is not resistant to chloride.

Comparing the corrosion resistances of C-276 and Alloy 59 as given, it is apparent that Alloy 59 has a better resistance to corrosion and is likely to last about 13 times longer than C-276 under similar conditions.

Alloy 59 should therefore, be recommended.

Answer is (C)

MB16:

Carbon steel cannot be used above 900 °F. The strength of aluminum declines greatly above 300 °F. Therefore, these two materials can be ruled out. Incoloy 800 has excellent corrosion resistance in both oxidizing and reducing atmospheres at high temperatures in absence of sulfur. However its cost is very high.
Austenitic steels (16 to 26 % Cr, 0.03 % max. carbon) are heat resistant, and because of their high chromium content are usually resistant to hydrogen atmosphere.

Answer is (C)

Rate data and interpretation:

MB17:

Solution of the rate equation for unimolecular reversible reaction is as follows

$$-\ln\left(1 - \frac{X_A}{X_{AE}}\right) = \frac{M}{M + X_{AE}} k_1 t$$

where X_A = conversion at time t
X_{AE} = equilibrium conversion
$M = C_{B0}/C_{A0}$

A plot of $-\ln\left(1 - \frac{X_A}{X_{AE}}\right)$ vs t should be a straight line whose

$$\text{slope} = \frac{M}{M + X_{AE}} k_1$$

The given plot of the data shows a straight line.

The slope of the line = (5.2 - 0)/(4.2 - 0) = 1.2381

M = 0.5/0.8 = 0.625

Therefore, $\frac{M+1}{M+X_{AE}} k_1 = 1.2381$

and then $\frac{0.625 + 1}{0.625 + 0.35} k_1 = 1.2381$

$$k_1 = \frac{1.2381}{1.667} = 0.7429 \text{ h}^{-1}$$

Answer is (B)

MB18:

Arrhenius relation: $k = a e^{-E/RT}$

Taking logarithm's of both sides gives $\ln k = \ln a - \frac{E}{R}\frac{1}{T}$

If ln k is plotted against 1/T, a straight line results. Its slope is $-\frac{E}{R}$

Calculations of ln k and 1/T are given below and plotted in the figure that follows.

1/T	3.3 x10⁻³	3.2 x10⁻³	3.1x10⁻³	3.0 x10⁻³	2.92x10⁻³
ln k	- 0.653	0.0953	0.79	1.3863	1.792

Plot these data as in figure below

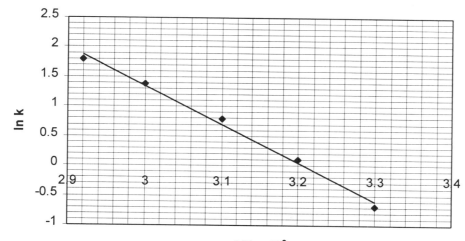

Plot of ln k vs 1/T

1/T x 10³

The slope of the line is $-\frac{E}{R} = -6.579$

Therefore, $E = 6.579 \times 10^3 \times 1.987 = \mathbf{13072}$ cal/gmol

Answer is (A)

MB19:

Rate equation in terms of conversion is $\dfrac{dX_A}{dt} = k_1(1 - X_A) - k_2 X_A$

At equilibrium, $\dfrac{dX_A}{dt} = 0$, $K = \dfrac{k_1}{k_2} = \dfrac{X_{AE}}{1 - X_{AE}}$ where K is equilibrium constant.

Solution of the equation in terms of X_{AE} is

$$-\ln\left(1 - \dfrac{X_A}{X_{AE}}\right) = \dfrac{1}{X_{AE}} k_1 t \qquad t = 8 \text{ min.}$$

Therefore, $-\ln\left(1 - \dfrac{0.333}{0.667}\right) = \dfrac{1}{0.667} k_1 (8)$

$k_1 = 0.0577$ min^{-1} $\qquad K = \dfrac{k_1}{k_2} = \dfrac{0.667}{1 - 0.667} = 2$

$k_2 = \dfrac{k_1}{K} = \dfrac{0.0577}{2} = 0.029$ min^{-1}

Answer is (B)

MB20:

For a reaction of nth order, $t_{½}$ half-life i.e at 50 % conversion is given by

$$t_{½} = \dfrac{2^{n-1} - 1}{k(n-1) C_{Ao}^{n-1}}$$

Taking logarithms of both sides, one obtains

$$\ln t_{1/2} = \ln \dfrac{2^{n-1}}{k(n-1)} - (n-1) \ln C_{Ao}$$

Thus a plot of $t_{1/2}$ vs $\ln C_{Ao}$ gives a straight line of slope $(1-n)$

The given data are to be plotted as in the following figure.

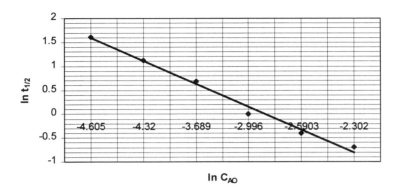

Plot of ln $t_{1/2}$ vs lnC_{AO}

The slope of the line = $\dfrac{1.6 - (-0.8)}{-4.605 - (-2.302)} = 1.0434 \cong -1.0$

Therefore, $1 - n = -1$ and **n = 2**

Answer is (B)

Reactor types:

MB21:

The rate of reaction in terms of conversion can be written as

$$-r_A = kC_A^2 = C_{Ao}\dfrac{dX_A}{dt} = C_{Ao}^2(1 - X_A)^2$$

Therefore, $\dfrac{dX_A}{dt} = kC_{Ao}(1 - X_A)^2$

For a mixed reactor, $\tau = \dfrac{V}{v_0} = \dfrac{C_{Ao}(1 - X_{Af})}{kC_{Ao}(1 - X_{Af})^2} = \dfrac{1}{k(1 - X_{Af})}$

where X_{Af} = concentration of effluent from the mixed reactor.

Now writing similar equation for the large reactor and taking the ratio, the following can be obtained

$$\dfrac{\tau_6}{\tau_1} = \dfrac{V_2}{V_1} = \dfrac{\frac{1}{1 - X_{Af2}}}{\frac{1}{1 - X_{Af1}}} = \dfrac{1 - X_{Af1}}{1 - X_{Af2}} = \dfrac{6}{1}$$

$1 - X_{Af2} = (1 - X_{Af1})/6 = \dfrac{1 - 0.5}{6} = 0.5/6$

$X_{Af2} = 1 - 0.5/6 = 0.917$

% conversion = 0.917 x 100 = 91.7 %

Answer is (B)

MB22:

For a plug flow reactor, space-time is given by

$$\tau = C_{Ao}\int_0^{X_A} \frac{dX_A}{kC_{Ao}(1-X_A)^2} = \frac{1}{k}\left[\frac{X_A}{1-X_A}\right]$$

Since all other things are the same, taking the ratio again, one gets

$$\frac{\tau_P}{\tau_m} = \frac{\frac{1}{k}\frac{X_A}{1-X_A}}{\frac{1}{k}\frac{1}{1-X_{Af1}}} = 1$$

since τ is the same for mixed and plug flow reactor

$$\frac{X_{Ap}}{1-X_{Ap}} = \frac{1}{1-X_{Af1}} = \frac{1}{1-0.5} = 2$$

From which, $X_{AP} = 0.67$ % conversion = 67 %

Answer is (A)

MB23:

min V = 7.66 x 30 $\cong$ 230 gal   Material balance on each reactor gives

relation between outlet concentration and inlet concentration. For first reactor,

$$C_1 = \frac{C_{Ao}}{(1+k_1\tau)}$$

$$C_2 = \frac{C_1}{(1+k_2\tau)} = \frac{C_{Ao}}{(1+k_1\tau)(1+k_2\tau)}$$

$$C_3 = \frac{C_2}{(1+k_3\tau)} = \frac{C_{AO}}{(1+k_1\tau)(1+k_2\tau)(1+k_3\tau)}$$

$$C_4 = \frac{C_3}{(1+k_4\tau)} = \frac{C_{Ao}}{(1+k_1\tau)(1+k_2\tau)(1+k_3\tau)(1+k_4\tau)}$$

$C_{ao} = 1.5$ lb moles/gal $C_4 = (1.5 - 1.5 \times 0.95) = 0.075$

$$(1+k_1\tau)(1+k_2\tau)(1+k_3\tau)(1+k_4\tau) = \frac{C_{Ao}}{C_4} = \frac{1.5}{0.075} = 20$$

Rate constants are known. The value of τ can be found by trial and error. Some calculations are given below

LHS product for different values of τ

Assumed	τ	6	8	7.66
LHS value		12.7	21.86	20.05 close

$$\frac{V}{v_0} = 7.66$$

V = 7.66(30) $\cong$ 230 gal

Answer is (C)

MB24:

If all the reactors are maintained at 15 °C, k_1 = 0.0806 min^{-1}

For this case, the exit concentration from the last reactor is given by

$$\frac{C_{Ao}}{C_A} = (1+k\tau)^N = \frac{1.5}{0.075} = 20, \quad N\ln(1+k\tau) = \ln(20)$$

$$N = \frac{\ln 20}{\ln[1+0.0806(7.66)]} = 6.23 \quad \text{Need more than 6 but less than 7.}$$

Since reactor number must be a whole number, 7 reactors should be used.

Answer is (D)

Economics:

MB25:

Cost of column including internals per tray =

$$\frac{96-90}{100-90} \times (2700-2250) + 2250 = \$2520$$

Cost of tower in 1998 = $\frac{390}{230} \times (2520)(30) = \$128,200$

Cost of instrumentation and piping = 0.6(128200) = $76800

Total fixed capital investment for the column = 128200 + 76800

= $205,000

Answer is (B)

MB26:

Condenser duty = 9.5x10^6 Btu/h

Operating hours per year = 8640 h

$$\text{water rate} = \frac{9.5 \times 10^6}{40(1)} = 237500 \text{ lb/h} = 474 \text{ gpm}$$

$$\text{Annual cost of cooling water} = \frac{474(60)(8640)(0.1)}{1000} = \$24,572$$

Answer is (C)

MB27:

Reboiler duty = 11.22 x 10⁶ Btu/h

From steam tables, latent heat of vaporization = 857.1 Btu/lb

$$\text{Steam usage per hour} = \frac{11.22 \times 10^6}{857.1} = 13091 \text{ lb/h}$$

$$\text{Annual cost of steam} = \frac{13091(8640)}{,1000} \times (2) = \$226,213$$

Answer is (A)

MB28:

Total fixed capital investment

Tower cost (calculated in MB25) = $205,000

Condenser cost = $\left[\frac{1270-1000}{1400-1000}\times(21150-17250)+17250\right]\times\frac{390}{230}\times 1.6 = \$53942$

Reboiler cost = $\left[\frac{2500-2200}{2600-2200}\times(30300-27750)+27750\right]\times\frac{390}{230}\times 1.6 = \$80476$

Pump101 cost = $\left[\frac{12250-10000}{20000-10000}(16000-11600)+11600\right]\times\frac{390}{230}\times 1.6 = \$34157$

Pump102 cost = $\left[\frac{3500-3000}{4000-3000}(8000-7200)+7200\right]\times\frac{390}{230}\times 1.6 = \$20619$

V101 cost = $\left[\frac{210-200}{300-200}(1900-1600)+1600\right]\times\frac{390}{230}\times 1.6 = \$4422$

Total fixed capital investment = $398616

Variable annual cost = 398616(0.15) + steam cost + cooling water cost

= 59792 + 226213 + 24572 = $310,577

Answer is (A)

THERMODYNAMICS

Compressors and expanders:

MB29:

Mass balance reduces to $\Sigma M_I = \Sigma M_o$

Energy balance reduces to $W = M_S(\bar{H}_I - \bar{H}_o)$ H = enthalpy, Btu/lb

Since process is adiabatic, Q = 0, $\bar{S}_I = \bar{S}_o$ S = entropy, Btu/lb.°F

At 150 psia and 900 °F, $\bar{S}_I = \frac{1.8301 + 1.8451}{2} = 1.8376$ Btu/lb.°F

$H_I = 1477.85$ Btu/lb

At 20 psia, $\bar{S}_o = 1.8376$ Btu/lb.°F

$$t = \frac{1.8376 - 1.7808}{1.8396 - 1.7808} \times (400 - 300) + 300 = 396.6 \text{ °F}$$

$$\bar{H}_o = \frac{396.6 - 300}{400 - 300} \times (1239.2 - 1191.6) + 1191.6 = 1237.97 \text{ Btu/lb}$$

Actual work = 0.85(1477.85 - 1237.97) = 203.9 Btu/lb

Therefore, turbine output = $\frac{20000 \, (203.9)}{2545} = 1602.4$ hp

Answer is (A)

MB30:

P = 100 psia,

The throttling process is isenthalpic. Therefore, $\bar{H}_I = 1477.85$ Btu/lb

$$t = \frac{1477.85 - 1428.9}{1479.5 - 1428.9} \times (900 - 800) + 800 = 896.7 \text{ °F}$$

$\bar{S}_I = 0.9674(1.8829 - 1.8443) + 1.8443 = 1.8816$ Btu/lb.°F

For maximum work, $\bar{S}_o = 1.8816$ Btu/lb.°F

$$t_o = \frac{1.8816 - 1.8396}{1.8918 - 1.8396} \times 100 + 400 = 480.5 \text{ °F}$$

$$\bar{H}_o = \frac{480.5 - 400}{500 - 400}(1286.6 - 1239.2) + 1239.2 = 1277.36 \text{ Btu/lb}$$

At 80 % efficiency actual work = 0.8(1477.85 - 1277.36) = 160.4 Btu/lb

$$\bar{H}_o = 1477.85 - 160.4 = 1317.45 \text{ Btu/lb}$$

$$t_o = \tfrac{1317.45 - 1286.6}{1334.4 - 1286.6}(600 - 500) + 500 = 565 \text{ °F}$$

Answer is (C)

MB31:

For minimum work, process is reversible, and isentropic

and $\bar{S}_I = \bar{S}_o = 1.4896$ Btu/lb.°R

Calculate outlet enthalpy and temperature.

From the properties at 60 psia,

$$t_o = 240 + \tfrac{1.4896 - 1.4819}{1.4976 - 1.4819}(20) = 249.8 \simeq 250 \text{ °F}$$

$$\bar{H}_o = 739.7 + \tfrac{1.4896 - 1.4819}{1.4976 - 1.4819}(750.9 - 739.7) = 745.2 \text{ Btu/lb of NH}_3$$

Therefore, isentropic reversible work is by energy balance,

$$W = \bar{H}_I - \bar{H}_o = 643.6 - 745.2 = -105.6 \text{ Btu/lb of ammonia.}$$

The negative sign indicates that work is done on the system.

If the compression efficiency is 80 %, the actual work will be

Actual work = - 105.6/0.8 = **- 132 Btu/lb of NH₃**

Answer is (D)

Mb32:

Actual work = $\bar{H}_I - \bar{H}_o$ = - 132 Btu/lb of ammonia calculated above.

Therefore $\bar{H}_o = \bar{H}_I + 132 = 775.6$ Btu/lb of NH₃

This value is not available from the table. Therefore, using C_p value (assumed constant),

$$t_o = 300 + \tfrac{775.6 - 773.3}{0.6} = 303.8 \cong 304 \text{ °F}$$

Entropy balance gives $\bar{S}_o - \bar{S}_I = \bar{S}_P$

Where $\bar{S}_P$ = Entropy increase during the irreversible process Btu/lb°R

At P = 60 psia and 300 °F $\bar{S} = 1.5281$ Btu/lb°R

Entropy change from t = 300 °F to 304 °F is given by

$$\Delta S = \int_{300+460}^{304+460} \frac{C_p dT}{T} = 0.6 \ln \frac{764}{760} = 0.00315 \text{ Btu/lb°R}$$

Therefore, actual entropy at outlet conditions = 1.5281 + 0.00315 = 1.5313 Btu/lb°F

Entropy increase in irreversible compression = 1.5313 − 1.4896 = 0.0417 Btu/lb°F

Answer is (B)

Estimation of thermodynamic properties:

MB33:

Calculate heat of vaporization at 25 °C = 273.1 + 25 = 298 K using Watson's relation

$$T_{r2} = \frac{298}{699} = 0.4263 \qquad T_{r1} = \frac{486.6}{699} = 0.6961$$

$$\Delta H_{V298} = \left(\frac{1 - 0.4263}{1 - 0.6961}\right)^{0.38} \times 11035 = 14049 \text{ cal/g mol}$$

Entropy change from saturated liquid to saturated vapor

$$\Delta S^{sat} = \frac{14049}{298} = 47.1 \text{ cal/(g mol.K)} = 47.1 \text{ (Btu/lb mol.R)}$$

= 47.1/150.177 = **0.314 Btu/(lb.R)**

Answer is (B)

MB34:

$$d\bar{G}_i = RT d\ln f_i$$

Starting with above equation, we get

$$RT \ln \frac{f}{f^*} = G - G^*$$

and since $\bar{G} = \bar{H} - T\bar{S}$

$$\ln \frac{f}{f^*} = \frac{1}{R}\left[\frac{\bar{H} - \bar{H}^*}{T} - (\bar{S} - \bar{S}^*)\right]$$

With reference state at low pressure, f* = P*.

$$\ln \frac{f}{P^*} = \frac{1}{R}\left[\frac{\bar{H} - \bar{H}^*}{T} - (\bar{S} - \bar{S}^*)\right]$$

Assume methyl chloride to behave as ideal gas at 6 psia

$$\ln \frac{f}{P*} = \frac{50.5}{1.987}\left[\frac{237.12 - 246.6}{240 + 460} - (0.4129 - 0.5607)\right]$$

$$\frac{f}{P*} = 30.3311$$

Fugacity $\quad f = 30.3311(6) = 182$ psia

Fugacity coefficient $\quad \phi = \frac{f}{P} = \frac{182}{200} = 0.91$

Answer is (A)

MB35:

$$\text{Saturated vapor} \rightarrow \textit{ideal gas state} \rightarrow \textit{compression} \rightarrow \textit{actual vapor state}$$

1 atm =14.7 psia	1 atm	228.3 psia	228.3 psia
t = 77 °F	77 °F to 190 °F	190 °F	190 °F
H = 150 Btu/lb		S_o = 0.325 Btu/lb.R	

$$\bar{S} = \bar{S}_o + \int_{460+77}^{460+190} \frac{C_P' dT}{MT} - \frac{R\ln \frac{P_2}{P_1}}{M} - \Delta S' \qquad M = \text{molecular wt}$$

$\Delta S' = -0.103 \; \textit{psia.ft}^3/\textit{lb.R} \times 144/778 = 0.019$ Btu/lb.R

$$\bar{S}_o = 0.325 + \int_{537}^{650} \frac{3.332 + 41.786 \times 10^{-3} T - 7.325 \times 10^{-6} T^2}{MT} dT - \frac{R\ln(228.3/14.7)}{M} - \Delta S'$$

$$= 0.325 + \left[\frac{3.332 \ln (650/537)}{58.12} + \frac{41.786 \times 10^{-3}(650-537)}{58.12} - \frac{3.6625 \times 10^{-6}(650^2 - 537^2)}{58.12} - \frac{1.987 \ln 1.9612}{58.12} - 0.019\right]$$

$$= 0.325 + 0.0837 - 0.0938 - 0.019$$

$$= \mathbf{0.2959 \; Btu/lb.R}$$

Answer is (C)

MB36:

Residual volume is the difference between the volume of ideal gas and the actual volume at the same conditions of temperature and pressure.

Actual volume per lbmol = 0.1296(64) = 8.2944 ft³/lb mol

Volume of SO_2 gas at the same conditions if it were an ideal gas at these conditions

$$V = \frac{RT}{P} = \frac{10.73(460+480)}{1000} = 10.0862 \; \text{ft}^3/\text{lb mol}$$

Residual volume = 10.0862 - 8.2944 = 1.7918 ft³/lb mol

$$= \frac{1.7918}{64} = 0.028 \text{ ft}^3/\text{lb}$$

Answer is (B)

V-L Composition of miscible and partially miscible systems:

MB37:

On the t-x diagram, the azeotropic point is shown as E. Reading the composition, the water mol fraction in the azeotrope is 0.67.

Therefore, mol fraction of i-butanol = 1 - 0.67 = 0.33

$$\text{Wt \% of i-Butanol in azeotrope} = \frac{0.33\ (71.14)}{0.33\ (71.14)+\ 0.67(18.02)} \times 100 = 66.04$$

Answer is (B)

MB37 to MB40: i-Butanol forms a minimum boiling-point azeotrope. The t-x diagram for this system is shown in the following figure.

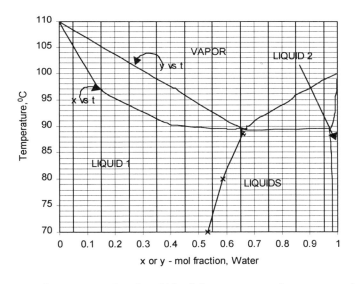

i-Butanol - Water system

MB37: i-Butanol concentration [wt %] of the azeotrope is most nearly

(A) 61.7 (B) 66.04 (C) 69.1 (D) 56.3

MB38:

$$\text{Mol fraction of water in liquid mixture } \frac{95/18.02}{95/18.02\ +\ 5/71.14} = 0.987$$

From the t-x diagram, the bubble point is found to be about 89.8 ≑ 90 ºC.

Answer is (B)

MB39:

From t-x diagram, at 80 ºC, the composition of butanol rich layer is

    mol fraction of water = 0.56
    mol fraction of butanol = 1 - 0.56 = 0.44

Therefore, wt % of i-Butanol = $\dfrac{0.44(71.14)}{0.44(71.44) + 0.56(18.02)} = 75.62$ wt %

Answer is (C)

MB40:

mol fraction of water in vapor = 1 - 0.15 = 0.85

Locate point (y_w = 0.85) and t = 110 on the equilibrium diagram, and move down vertically to meet y vs t curve. At the intersection point, read temperature to the right or left. The temperature is 92.8 ºC approximately

Answer is (C)

END OF SOLUTIONS OF PM EXAM 3..

SOLUTIONS TO PM SAMPLE EXAM 4
FLUIDS
Properties of fluids:

AB1:

T_c = 273.15 + 320 = 593.15 K T/T_c = 0.7
Therefore, T = 0.7(693.15) = 415.2 k = 142 °C
Vapor pressure of toluene at 142 °C by interpolation,

$$P_v = \frac{142 - 136.5}{178 - 136.5}(3 - 2) + 2 = 2.1325 \text{ atm.}$$

$$\omega = -\log_{10} P_r - 1 = -\log(\frac{2.135}{41.6}) - 1 = 0.29$$

Answer is (D)

AB2:

Watson's equation: $\Delta H_{v2} = \Delta H_{v1}\left(\frac{1 - T_{r2}}{1 - T_{r1}}\right)^{0.38}$

$$T_{r1} = \frac{273.15 + 97.2}{273.15 + 263.7} = \frac{370.35}{536.85} = 0.690$$

$$T_{r2} = \frac{273.15 + 115}{273.15 + 263.7} = \frac{388.15}{536.85} = 0.723$$

$$\Delta H_{v2} = 164.36\left[\frac{1 - 0.723}{1 - 0.69}\right]^{0.38} = 157.5 \text{ cal/g}$$

$$= 157.5(4.1868)(60.09)$$

$$= \textbf{39625} \text{ J/gmol}$$

Answer is (D)

AB3:

From steam tables, at 70 °F, h_L = 38.04 Btu/lb

Again from steam tables, at 260 psia and 1500 °F,

$$H_v = \frac{1579.1 + 1632.5}{2} = 1605.8 \text{ Btu/lb}$$

Enthalpy change = 10 (1605.8 - 38.04) = **15677.6** Btu

Answer is (B)

171

AB4:

Concentration of NaOH: Basis of calculation 100 g of solution
gmol of NaOH = 10/40 = 0.25 gmol

90 g of water = 90 cc with sp gr = 1.0
C_i = (0.25/90)(1000) = 2.7778 gmol/Liter

(assumes no volume change on dissolving NaOH in water)

k of water at 20 °C = 0.340 Btu/(h.ft.°F) = $\dfrac{0.34 \times 252 \times 1.8}{3600 \times 30.48}$ = 0.0014055 $\dfrac{cal}{cm.s.K}$

k_{mix} = 0.0014055 + $\dfrac{1}{4.186}$ Σ(2.7778 × 0 + 2.7778×20.934×10^{-5})

= 0.0015444 cal/(cm.s.K) = **0.3736** Btu/(h.ft.°F)

Answer is (A)

KINETICS

Control of reactors:

AB5:

Cascade temperature control alone is good to overcome the time lags inherent in a reactor system. However, when a batch is required to attain a certain temperature to initiate the reaction, the heating and cooling medium control valves are split range controlled. The heating medium control valve operates between air signal values of 9 and 15 psig and the cooling medium control valve operates between 3 and 9 psig. In the event of instrument air failure, the heating medium valve closes and the cooling medium valve opens to provide emergency cooling. Answer is split range control.

Answer is (B)

AB6:

The transfer function for set point changes can be written as

$$C_2(s) = \dfrac{0.12 K_c \left(\dfrac{1}{Ts+1}\right)^2}{1 + 0.12 K_c \left(\dfrac{1}{Ts+1}\right)^2 e^{-0.5s}} R(s)$$

Offset = $C_2(\infty) - R(\infty)$

$R(\infty) = 1$ *the desired value.* Since forcing function is unit step.

Steady state value of the response can be obtained by final value theorem as

$$C_2(\infty) = \lim_{s \to 0} [sf(s)] = \lim_{s \to 0} s \cdot \frac{0.12K_c \left(\frac{1}{Ts+1}\right)^2}{1 + 0.12K_c \left(\frac{1}{Ts+1}\right)^2 e^{-0.5s}} = \frac{0.12K_C}{1 + 0.12K_C}$$

Therefore, offset $= 1 - \frac{0.12K_C}{1 + 0.12K_C} = \frac{1}{1 + 0.12K_C}$

Answer is (A)

AB7:

Calculate k at conversion of 0.5

$$\frac{x}{1-x} = \frac{kV}{F} = k(1.2) \qquad k = \frac{1-0.5}{0.5(1.2)} = 0.833 \ h^{-1}$$

Now calculate temperature at this k.

$$\ln k - \ln k_{150} = -\frac{E}{R}\left[\frac{1}{T} - \frac{1}{T_1}\right] = -\frac{30000}{1.987}\left[\frac{1}{T} - \frac{1}{610}\right]$$

From which T = 599 R = 139 °F

Critical temperature difference = $\Delta T_c = \frac{RT^2}{E} = \frac{1.987(599)^2}{30000} = 23.8 \ °F$

Answer is (C)

AB8:

At the intersection of heat generation and heat removal curves, T = 139 °F
Apply stability criteria as follows

From graph, Q = 300000 Btu/h at the intersection point.

$$\Delta T_c = 23.8 \ °F \qquad \left(\frac{\partial Q}{\partial T}\right)_x = \frac{300000}{23.8} = 12605 \ \text{Btu/°F}$$

$$UA + F\rho C_P\left(2 + \frac{kV}{F}\right) = 50 \times 150 + 2000 \times 0.75(2 + 0.833 \times 1.2) = 11999 \ \text{Btu/°F}$$

$$UA + F\rho C_P\left(2 + \frac{kV}{F}\right) \not> \left(\frac{\delta Q}{\delta T}\right)_x$$

So the most important criterion for stability is not met.
Hence the reactor will be unstable with a constant jacket temperature.

[With slight temperature rise, a still higher temperature will be obtained but since there is no upper stable point (see figure), the temperature will come down after the the reactant concentration has decreased somewhat. Also, lowering the temperature of feed to (70-50) = 20 °F will not make the operation stable as both UA and $F\rho C_P$ have not changed. With feed at 20 °F, all the heat can be removed giving a lower critical temperature difference as calculated below

Sensible heat out with product = 2000(0.75)(139-70) = 104000 Btu/h

Heat removed through the jacket = 300000 - 104000 = 196000 Btu/h

$$T - T_j = \frac{196000 - 2000 \times .75 \times (70-20)}{50 \times 150} = 16.1 \text{ °F} < \Delta T_c = 23.8$$

This criteria alone is not sufficient to determine stability.
The reactor operation can be made stable by controlling the jacket temperature with a feedback controller.]

Answer is (D)

Endothermic/exothermic reactors:

AB9:

From the optimal progression curve provided in the problem, at $X_A = 0.8$, read reaction temperature = 62 °C = 273 + 62 = 335 K.
Calculate reaction rate constants:

$$k_1 = e^{17.2 - 11600/RT} = e^{17.2 - 11600/(1.987 \times 335)} = 0.7972 \text{ min}^{-1}$$

$$k_1 = e^{41.9 - 29600/RT} = e^{41.9 - 29600/(1.987 \times 335)} = 0.766775 \text{ min}^{-1}$$

Calculate rate of reaction

$$-r_A = C_{AO}[k_1(1 - X_A) - k_2 X_A]$$

$$= 4[0.7972(1 - 0.8) - 0.07668(0.8)] = 0.392 \quad \text{gmol/min}$$

Volume of reactor = $\frac{F_{AO} X_A}{-r_A} = \frac{980(0.8)}{0.392} = 2000$ liters

Answer is (B)

AB10:

Choosing T_1 as datum temperature, the heat balance can be written as follows

$$[C_p(T_2 - T_1)X_A + C_p(T_2 - T_1)(1 - X_A)] + \Delta H_r X_A = 0$$

Which reduces to $\quad X_A = \dfrac{C_p \Delta T}{-\Delta H_R}$

This is energy balance equation. Its slope can be calculated as

$$\dfrac{C_p}{-\Delta H_r} = \dfrac{1000 \; cal/4 \; gmol \; A.K}{18000 \; cal/mol \; A} = \dfrac{1}{72} \; K^{-1}$$

Answer is (A)

AB11: Plot this energy balance line on optimal conversion diagram as shown below

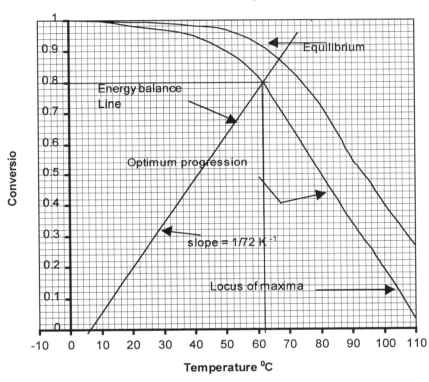

Equilibrium and optimal progression conversions as a function of temperature

The energy balance line meets the temperature axis at t = 5 °C approximately
Cooling of feed :   (250 cal/mol)(25 - 5) (980 mol) = 4.9 x 10⁶ cal/min⁻¹
Cooling of product = (250 cal/mol)(62 - 25) (980 mol) = 9.065 x 10⁶ cal/min⁻¹
Total cooling duty = (4.9 x 10⁶ + 9.065 x 10⁶) = **13.965x10⁶ cal/min**

Answer is (A)

AB12:

The minimum space-time for a plug flow reactor is given by

$$\tau = C_{Ao} \int \frac{dX_A}{(-r_A)_{opt}}$$

To determine τ, plot $(1/-r_A)$ vs X_A and determine the area under the curve or evaluate the area by Simpson's rule.

X_A	0.3	0.5	0.6	0.7	0.8
$-\frac{1}{r_A}$	0.17	0.42	0.68	1.205	2.5

Plot of 1/-r_A vs conversion, X_A

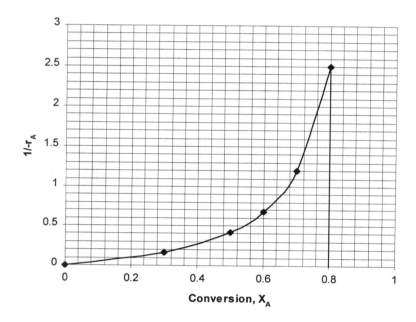

The plot of $-\frac{1}{r_A}$ vs X_A is made in the figure. The area under the curve by counting the squares is $105(0.1)(0.04) = 0.42$

[By Simpson's rule, area under the curve

$$= \frac{0.2}{3}[0 + 4 \times 0.1 + 2 \times 0.27 + 4 \times 0.7 + 2.5] = 0.416$$

$$\tau = C_{Ao}(0.416) = 4(0.416) = 1.66 \text{ min.}$$

Answer is (A)

Product distribution:

AB13:

For reactions in parallel, concentration levels of reactants is the key to proper control, the product distribution. Divide fiest equation by the second to obtain

$$\frac{r_s}{r_R} = \frac{k_2}{k_1} C_A^{-0.5} C_B^{1.6}$$

This ratio has to be as low as possible to favor first reaction. For this C_A should be high and C_B should be low. Also the concentration dependency of B is more pronounced than that of A, it is more important to have less B than high C_A. Therefore, method A is to be used.

Answer is (A)

AB14:

$$C_B = C_{A0} \frac{k_1}{k_2 - k_1}(e^{-k_1 t} - e^{-k_2 t})$$

Differentiating for a maximum the above equation gives

$$\frac{dC_B}{dt} = C_{A0} \frac{k_1}{k_2 - k_1}(-k_1 e^{-k_1 t} + k_2 e^{-k_2 t}) = 0$$

From which, $t = \frac{\ln(k_1/k_2)}{k_1 - k_2} = \frac{\ln(0.35/0.13)}{0.35 - 0.13} = 4.55$ h

Then $C_{B\,max} = \frac{0.35\,(5)}{0.13 - 0.35}(e^{-0.35(4.55)} - e^{-0.13\,(4.55)})$

$= \mathbf{2.79}$ lb mole/ft³

Answer is (C)

AB15:

In case of single continuous mixed reactor, differentiating the equation for C_{B1} gives

$$\frac{dC_{B1}}{dt} = (1 + k_1\theta)(1 + k_2\theta) - \theta[k_1(1 + k_2\theta) + k_2(1 + k_1\theta)] = 0$$

From which, $\theta = \sqrt{\frac{1}{k_1 k_2}} = \sqrt{\frac{1}{(0.35)(0.13)}} = 4.7$ h

$$C_{B1} = \frac{(0.35)(5)(4.7)}{[1 + (0.35)(4.7)][1 + (0.13)(4.7)]} = \mathbf{1.925} \text{ lb mole/ft}^3$$

Answer is (C)

AB16:

For first order parallel reactions, the rate equations are

$$-\frac{dn_a}{dt} = (k_1 + k_2 + k_3)n_a = k\, n_a$$

$$\frac{dn_b}{dt} = k_1 n_a \qquad \frac{dn_c}{dt} = k_2 n_a \qquad \frac{dn_d}{dt} = k_3 n_a$$

Solution of first equation is $n_a = n_{Ao} e^{-kt}$ by direct integration

Substituting this relation in other equations and integrating gives the following

$$n_b = n_{bo} + \frac{k_1}{k} n_{Ao}(1 - e^{-kt})$$

$$n_c = n_{co} + \frac{k_2}{k} n_{Ao}(1 - e^{-kt})$$

$$n_d = n_{do} + \frac{k_3}{k} n_{Ao}(1 - e^{-kt})$$

After one hour of reaction,

$$n_d = 0 + \frac{0.1}{0.36 + 0.12 + 0.1} \times 100 [1 - e^{(0.36 + 0.12 + 0.1)(1)}]$$

$\doteq 7.59$ gmols of D

Answer is (A)

Reactor design:

AB17:

For a batch reactor, $t = - \int_{C_{Ao}}^{C_A} \frac{dC_A}{-r_A}$

Thus the integral $- \int_{C_{Ao}}^{C_A} \frac{dC_A}{-r_A}$ will directly give the batch processing time.

The integral can be obtained by plotting $-1/r_A$ vs C_A and getting the area under the curve by counting the squares or by Simpson rule.
[To apply Simpson rule or trapezoidal rule, it is not absolutely necessary to have the plot.]

Obtain the area under the curve as follows

Apply trapezoidal rule over $C_A = 0.05$ And $C_A = 0.1$ lb mol/ft³

Area = $\frac{0.025}{2}[25 + \frac{25 + 12.3457}{2} + 12.3457]$ = 0.7

Area for the concentration range from 0.1 to 0.5 is calculated by Sinpson's rule as follows

Area = $\frac{0.1}{3}[1.1765 + 12.3457 + 4(1.8868 + 5.5556) + 2(3.2258)]$ = 1.66 ≐ 1.7

Total area = 2.4 which also is the time in hours for each batch.

Therefore number of batches = 24/2.4 = **10** batches per day.

Answer is (D)

AB18:

For a continuous mixed flow reactor, space time is given by

$$\tau = \frac{V}{v_0} = \frac{C_{Ao} - C_A}{-(r_{Af})} = \frac{0.5 - 0.05}{0.04} = 11.25 \text{ h}$$

Therefore volume required = 11.25 (25) = **281.3** ft³

Answer is (B)

AB19:

For two reactors in series each of volume 50 ft³, the space time is given by

$$\tau = \frac{50}{25} = 2 \text{ hours for each vessel}$$

For two reactors, $\tau_2 = n\tau_1 = 2(2) = 4$ hours = $\frac{n}{k}\left[\left(\frac{C_{Ao}}{C_{A2}}\right)^{\frac{1}{2}} - 1\right]$

From problem KH2-2, C_{A1} = 0.05 lb moles/ft³ and $\tau_1 = 11.25$ hours

and $11.25 = \frac{1}{k}\left[\left(\frac{C_{A1}}{C_{Ao}}\right) - 1\right]$

If we take the ratio of the above 2, k is eliminated and we can get

$$\frac{4}{11.25} = \frac{2}{1} \frac{\left[\left(\frac{0.5}{C_{A2}}\right)^{1/2} - 1\right]}{\left[\left(\frac{0.5}{0.05}\right) - 1\right]}$$

$$\left[\left(\frac{0.5}{C_{A2}}\right)^{1/2} - 1\right] = \frac{4}{11.25(2)}\left[\left(\frac{0.5}{0.05}\right) - 1\right]$$

179

From which $C_{A2} = 0.074$

$$\text{Percent conversion} = \frac{0.5 - 0.074}{0.5} \times 100 = 85.2\%$$

Answer is (C)

AB20:

For a plug flow reactor, $\tau = \frac{V}{v_o} = -\int_{C_{Ao}}^{C_{Af}} \frac{dC_A}{C_{Ao}}$

The integral is the same value as found in problem KH2-1 for t and is 2.4 hours

Volume of plug flow reactor = 25(2.4) = **60 ft³**

Answer is (A)

MASS TRANSFER

Phase diagrams:

AB21:

Use the given phase diagram to solve the problem. The construction of the solution is shown in the following figure.

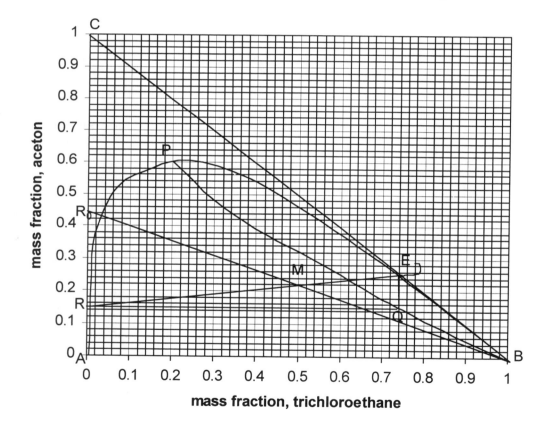

Phase diagram for acetone-water-trichloroethane system

Locate point R on the binodal curve with the help of the given raffinate composition. From R draw a line parallel to X axis and to meet the conjugate curve in Q. From Q, draw a vertical to meet binodal curve in E.

Read composition of extract from point E as $y_{acetone}$ = **0.254**,
Mass fraction of solvent = 0.74 and
Mass fraction of water by difference = 1 - 0.254 - 0.74 = 0.006

Answer is (B)

AB22:

Amount of extract:

$$R + E = 1000$$

By solvent balance,

$$0.007R + 0.74E = 500$$

Solution of the two equations gives R = 672.5 lb and E = 327.5 lb

Extract amount is **672.5** lb

Answer is (A)

AB23:

The mixture point of R and E lies on RE. It also lies on the vertical from point x_s = 0.5 since R_0 = S = 500 lb.(Application of lever rule).

Therefore, join B (solvent) and M located as explained above and extend to meet the Y axis in R_0. (R_0 lies on Y axis because solvent content of original mixture is zero)

R_0 gives acetone concentration = **0.44 mass fraction**

Answer is (C)

AB24:

Selectivity of solvent for solute acetone = $\beta = \dfrac{y_A/y_B}{x_A/x_B} = \dfrac{0.254/0.15}{0.006/0.843} = 277.9$

Answer is (B)

THERMODYNAMICS

Combustion:
AB25:

Basis: 1 lb of solid waste

component	lb	lb mole	Oxygen required for complete combustion, lb mole		Product of combustion, lb mole
H_2O	0.28	0.0156	---		
C	0.234	0.0195	0.0195	CO_2	0.0195
H	0.03	0.015	0.0075	H_2O	0.015
O	0.2	0.00625	----		
Ash	0.256	---	----		

Heat of combustion of garbage can be found from the heats of formation of CO_O and water.

ΔH_f of CO_O = - 94.05 kcal/g mol = - 94.05(1800) = - 169290 Btu/lb mol
ΔH_f of H_2O = - 68.314 kcal/g mol = - 68.314(1800) = - 122695.2 Btu/lb mol

$\Delta H_c = 0.0195(-169290) + 0.015(-122695.2) = 5141.6$ Btu/lb garbage.

Answer is (D)

AB26:

Energy balance on incinerator reduces to $\Sigma H_i = \Sigma H_o$
Use C_p and ΔH_f *data* to obtain enthalpies of the components. Prepare table as follows
Assume enthalpies of H, O and N to be 0 at 77 °F. Also enthalpy of ash = 0 at 77 °F.

component	enthalpy at 77 °F Btu/lb mol	enthalpy at 2000 °F Btu/lb mol
H_2	0	$n_H C_{p\,H2}(2000 - 77)$
O_2	0	$n_{CO2} C_{pCO2} (2000 - 77)$
N_2	0	$n_{N\,2} C_{p\,N2} (2000 - 77)$
$H_2O(l)$	- 122965.2	-----
$H_2O(v)$	-104040	$- 104040 + n_{H2O} C_{pH2O} (2000 - 77)$
CO_2	-169290	$- 169290 + n_{CO2} C_{pCO2} (2000 - 77)$

Oxygen requirements:
    For combustion of carbon = 0.0195 lb mole
    For combustion of H = 0.0075 lb mole
    Total for combustion = 0.027 lb mole
    O_2 available from garbage = 0.00625 lb mole
    Net O_2 required = 0.02075 lb mole
    Let excess O_2 be = x lb mole
    O_2 in flue gas = x lb mole
    N_2 in flue gas = (0.79/0.21)(0.02075 + x) lb mole

By energy balance,

$0.0156(-122965.2)$ = $0.0306(-104040) + 0.0306(9.339)(2000 - 77)$ water vapor
(water in garbage) $+ 0.0195(-169290) + 0.0195(12.01)(2000 - 77)$ CO_2
                                $+ x(7.976)(2000 - 77)$ O_2
                                $+ (0.07806 + 3.7619x)(7.452)(2000 - 77)$ N_2
                                $+ 0.3(0.256)(2000 - 77)$ Ash

solving for x,
       Excess oxygen, $x = 0.03322$ lb mole
       and nitrogen in air   = $(0.03322 + 0.02075)(3.7619) = 0.203$ lb mole

% excess air = $\frac{0.03322}{0.02075} \times 100 = 160\%$

Answer is (A)

AB27:

Products of combustion :

     Total water = $0.0156 + 0.015 = 0.0306$ lb moles
     CO_2 = 0.0195   = 0.0195 lb moles
     N_2 = 0.203   = 0.203   lb moles
     O_2 = 0.03322   = 0.03322 lb mole
               Total   = 0.28632 lb mole per lb of waste

Answer is (C)

AB28:

Component	lb mole	Enthalpy at 2000 °F	
H_2O (v)	0.0306	$0.0306[-104040 + 9.339(2000 - 77)]$	= – 2634.1
CO_2	0.0195	$0.0195[-169290 + 12.01(2000 - 77)]$	= – 2850.8
N_2	0.203	$0.2030[\quad\quad 7.452(2000 - 77)]$	= + 2911.9
O_2	0.03322	$0.03322[\quad\quad 7.976(2000 - 77)]$	= + 509.5
		Total	= –2063.5 Btu/lb waste

Component	lb mole	Enthalpy at 700 °F	
H_2O (v)	0.0306	$0.0306[-104040 + 8.341(700 - 77)]$	= –3024.6
CO_2	0.0195	$0.0195[-169290 + 10.262(700 - 77)]$	= –3176.5
N_2	0.203	$0.2030[\quad\quad 7.055(700 - 77)]$	= + 892.2
O_2	0.03322	$0.03322[\quad\quad 7.347(700 - 77)]$	= + 152.1
		Total	= –5156.2 Btu/waste

$-\Delta H = -2063.5 - (-5156.2) = 3092.7$ Btu/lb of waste

Heat required to produce 1 lb of steam at 500 °F and 165 psia

$$= H_v - h_l = 1272.6 - 48 \quad \text{from steam tables}$$
$$= 1224.6 \text{ Btu per lb of steam}$$

Steam production/lb of garbage = $\frac{3092.7}{1224.6} = 2.525$ lb/lb garbage

Answer is (A)

Economics:

AB29:

Select the insulation on the basis of incremental % returns.
Tabulate calculations as follows

Insulation Thickness in	1	2	3	4
Annual savings $	1,037	1,210	1,279	1,313
Fixed charges $	173	230	259	269
Net income $	864	980	1,020	1,044
% Return	60	51	47.2	46.5

Comparing 1 and 2, incremental % return = $\frac{980 - 864}{1920 - 1440} \times 100 = 24.17\%$

Comparing 2 and 3, incremental % return = $\frac{1020 - 980}{2160 - 1920} \times 100 = 16.8\%$

Comparing 3 and 4, incremental % return = $\frac{1044 - 1020}{2244 - 2160} \times 100 = 28.57\%$

By incremental % return method, 2 in is better than 1 in
3 in is better than 2 in
4 in is better than 3 in

Select 4 in insulation thickness.

Answer is (D)

AB30:

Depreciation = (12000 - 12000)/5 = $21600 yr based on straight line method

Net profit in first year = 36000 - 21600 = $ 14400
Net profit in second year = 37200 - 21600 = $ 15600
Net profit in third year = 43200 - 21600 = $ 21600
Net profit in fourth year = 48000 - 21600 = $ 26400
Net profit in fifth year = 51600 - 21600 = $ 30000
Time average value based on future worth :

$$14400(1+i)^4 + 15600(1+i)^3 + 21600(1+i)^2 + 26400(1+i) + 30000$$

$$= A(F/A, i = 0.15, 5) = A \frac{(1+i)^5 - 1}{i}$$

Or $14400(1.15)^4 + 15600(1.15)^3 + 21600(1.15)^2 + 26400(1.15) + 30000 = A \left(\frac{1.15^5 - 1}{0.15} \right)$

A = $20443.42

Answer is (C)

AB31:

$$\text{Payout period} = \frac{\text{depreciable fixed capital investment}}{\text{average profit/yr} + \text{depreciation/yr}}$$

Prepare a table as follows for convenient calculations

	investment 1	investment 2	investment 3
Depreciable amount $	99000	170000	209000
Annual cash flow $	39600	57200	65000
Depreciation $	19800	34000	41800
Net profit $	19800	23200	23200
Payout period yrs	2.5	2.97	3.22

Answer is (A)

AB32:

Break-even point occurs when total annual product cost = annual sales.

$$\text{direct production cost /unit} = \frac{350000}{700000 / 50} = \$25$$

At break even point, $250000 + 25N = 50N$
Solving for N,
$$N = 250000/(50 - 25) = 10000 \text{ units/yr}$$

Answer is (C)

Thermodynamic charts and data bases:

AB33:

Use P-H diagram.
At P = 150 psia, t = 100 °F, from chart saturated liquid properties are

$\bar{h} = 46$ Btu/lb $\qquad \bar{S} = 0.091$ Btu/(lb°R)

For maximum work output, entropy change is 0. Following constant entropy line down to 0 °F temperature, quality of turbine discharge is 33 % i.e. vapor amount is 33 % and liquid is 67 %.

Get values of enthalpy of saturated vapor and liquid at 100 °F from chart.

$\bar{H}_v = 103$ Btu/lb $h_l = 12.5$ Btu/lb

Enthalpy at turbine exit = 0.33(103) + 0.67(12.5) = 42.4 Btu/lb

Maximum work = 42.4 - 46 = - 3.6 Btu/lb

Actual work = - 3.6×0.85 = - 3.06 Btu/lb

Actual enthalpy at turbine exit = 46 - 3.06 = 42.94 $\cong 43$ Btu/lb

Reading the value of entropy at 100 °F and h = 43 Btu/lb,
$\bar{S} \cong 0.095$ Btu/(lb°R) approximately

Entropy change = 0.095 - 0.091 = 0.004 Btu/(lb°R)

Answer is (D)

AB34:

This problem requires the use of enthalpy-concentration diagram for NaOH-H$_2$O system.

First carry out the mixing adiabatically and determine the final temperature of the mixture. For this locate the two points representing the two solutions on the chart and read the point of intersection of straight line joining these two points and the 45 % concentration vertical. At this point, temperature isotherm indicates t = 210 °F and the enthalpy of mixed solution is **210 Btu/lb** of solution.

Now carry out cooling along 45 % composition line and going downwards read enthalpy of mixed solution = **95 Btu/lb**.

Heat to be removed = 95 - 210 = **-115 Btu/lb of solution**.

(Note: The sign of Q is negative because heat is removed from the solution.)

Answer is (A)

AB35:

At inlet, $P_{r1} = \frac{15}{48.2} = 0.3115$ $T_{r1} = \frac{460 + 90}{305.4 \times 1.8} \cong 1.00$

At outlet, $P_{r2} = \frac{96}{48.2} \cong 2.0$ $T_{r2} = \frac{460 + 145}{305.4 \times 1.8} = 1.1$

From the enthalpy generalized chart,

At inlet, $\dfrac{h_1^* - h_1}{RT_C} = 0.38$ and at outlet, $\dfrac{h_2^* - h_2}{RT_C} = 2.76$

$h_1^* - h_1 = 0.38(1.987)(305.4 \times 1.8) = 415.1$ Btu/lb mol

$h_2^* - h_2 = 2.76(1.987)(305.4 \times 1.8) = 3014.7$ Btu/lb mol

$h_2^* - h_1^* = \int_{460+90}^{460+145} (4.01 + 0.01636T)dT$

$\quad = 4.01(605 - 550) + 0.00818(605^2 - 550^2) = 740.2$ Btu/lb mol

W = $W = 415.1 + 740.2 - 3014.7 = -1859.4$ Btu/lb mol = **- 62 Btu/lb**

Answer is B

AB36:

Here $S_1 - S_2 = S_1^* - S_2 = S_1^* - S_2^* + S_2^* - S_2$

At outlet, $P_{r2} = \dfrac{96}{48.2} \doteq 2.0$ $\qquad T_{r2} = \dfrac{460+145}{305.4 \times 1.8} = 1.1$

$S_2^* - S_1^* = \int_{550}^{605} \dfrac{4.01 + 0.01636T}{T}dT - R \ln \dfrac{96}{15}$

$\quad = 4.01 \ln \dfrac{605}{550} + 0.01636(605 - 550) - 1.987 \ln \dfrac{96}{15}$

$\quad = 0.3822 + 0.8998 - 3.6685 = -2.4065$ Btu/(lbmol.°R)

At outlet, $\dfrac{S_2^* - S_2}{R} = 2.0$ from the generalized entropy chart

$S_2^* - S_2 = 1.987(2) = 3.974$ Btu/(lb mol. °R)

Entropy change = 3.9740 - 2.4065 = 1.5675 Btu/(lb mol. °R) = **0.05225 Btu/(lb.°R)**

Answer is (C)

Thermodynamic laws:

AB37:

Mixing effects are to be ignored. Assume t = °C as reference temperature.
An energy balance on the vessel as the system reduces to

$$0 = \Sigma H_i - \Sigma H_o + Q$$

where ΣH_i = total enthalpy coming in
ΣH_o = total enthalpy leaving the system
Q = heat loss in consistent units

Specific heats are in cal/(g mol.°C). So change the loss to cal/s from J/s

$$Q = (0.239)[209(t-25)] = 49.951(t-25) \text{ cal/sec}$$

Since t_{ref} = °C, at a temperature t, enthalpies for the two components can be written as

$$\bar{H}_{iO_2} = \int_0^t C_p dT = \int_0^t (6.9963 + 3.1 \times 10^{-3} t) = 6.9963t + 1.55 \times 10^{-3} t^2$$

$$\bar{H}_{iN_2} = \int_0^t C_p dT = \int_0^t (6.8413 + 1.25 \times 10^{-3} t) = 6.8413t + 0.625 \times 10^{-3} t^2$$

Inlet total enthalpy:
Enthalpy of O_2 at 100 °C = $10[6.9963(100) + 1.55 \times 10^{-3}(100^2)]$ = 7306.3 cal/10 mols
Enthalpy of N_2 at 150 °C = $10[6.8413(100) + 0.625 \times 10^{-3}(100^2)]$ = 6903.8 cal/10 mols

Therefore, $\Sigma H_i = 7306.3 + 6903.8$ cal/s = 14210.1 cal/s

Outlet total enthalpy
O_2 H = 10 [6.9963(t_o) + 1.550x10^{-3}(t_o^2)]
N_2 H = 10 [6.8413(t_o) + 0.625x10^{-3}(t_o^2)]
Total enthalpy out with gases = 138.376 (t_o) + 0.02175(t_o^2)

substituting in heat balance equation,

$$0 = 14210.1 - [138.376(t_o) + 0.02175(t_o^2)] - 49.951(t_o - 25)$$

which simplifies to

$$0 = 15458.875 - 188.327 t_o - 0.02175 \, t_o^2$$

$$t_o^2 + 8658.7126 t_o - 710752.8735 = 0$$

Solving the quadratic, t_o = 81.32 °C

Answer is (B)

(In exam, all the details and steps given above are unnecessary.)

AB38:

$$\Delta S = \int_{460+68}^{460+94.3} \frac{0.521 dT}{T} + \frac{151.06}{460+94.3} + \int_{460+94.3}^{460+122} \frac{0.44 dT}{T} \quad \text{at constant pressure}$$

$$= 0.521 \ln \frac{554.3}{528} + \frac{151.06}{554.3} + 0.44 \ln \frac{582}{554.3}$$

$$= 0.3193 \text{ Btu/lb.°R}$$

Therefore, entropy change per lb mol = 0.3193(74.12) = **23.67** Btu/(lb mol. °R)

Answer is (D)

AB39:

Energy balance for the conditions of the problem with vessel as system reduces to

$$U_E - U_B = Q$$

Steam tables do not give U values for superheated steam. By definition,

$$H = U + PV$$

Therefore $\Delta U = \Delta H - \Delta(PV)$

From steam tables, properties are

Initial condition	Final condition
150 psia	500 psia
t = 358.42 °F	t = 700 °F
$\bar{V}_{\ell 1} = 0.01809$ ft³/lb	$\bar{V}_{\ell 2} = 0$ ft³/lb. no liquid present
$V_{g1} = 3.015$ ft³/lb	$V_{g2} = 1.3044$ ft³/lb
$\bar{H}_v = 1194.1$ Btu/lb	$\bar{H}_v = 1357.0$ Btu/lb
$\bar{H}_l = 330.51$ Btu/lb	no liquid present. superheated vapor

Initial volume = final volume = 1.3044 ft³

Let x be the quality of steam initially. Then

$V_1 = 3.015x + 0.01809(1 - x) = 1.3044$ hence x = 0.4292

$\bar{H}_B = 1194.1(0.4292) + 330.51(1 - 0.4292) = 701.2 Btu/lb$

Q (heat added) = $(1357 - 701.2) - \frac{(500 - 150)(1.3044)(144)}{778}$

= 1357 - 701.2 - 84.5 = **571.3** Btu/lb

Answer is (C)

AB40:

By second law of thermodynamics,

$$\frac{W}{Q_H} = \frac{T_H - T_L}{T_H} = \frac{535 - 490}{535} = 0.08411$$

W = 0.08411(200000) = 16822 Btu/h

$$Q_L = \frac{W}{\frac{T_H}{T_L} - 1} = \frac{16822}{\frac{535}{490} - 1} = 183173 \text{ Btu/h}$$

Answer is (A)

END OF SOLUTIONS OF PM EXAM 4..

SOLUTIONS TO PM BONUS EXAM.
Equipment design:

B1:

Schedule 40 is quite adequate for the conditions of pressure and temperature specified. Pipe size to be found by trial. Assume 6 in std pipe size. D_i = 0.5052 ft, A_c = 0.2006 ft². For commercial steel pipe, surface roughness = 0.00015 ft ,

ε/D_i = 0.00015/0.5052 ≐ 0.0003

Specific gravity of liquid = 75/62.4 = 1.2, Viscosity = 0.28/1.2 = 0.233 cP

Vol. flow rate = $\frac{800}{60(7.48)}$ = 1.783 ft³/s, u = 1.783/0.2006 = 8.9 ft/s o.k.

Re. No. = $\frac{0.5052(8.9)(46.4)}{0.23(0.000672)}$ = 1.35×10^5 f = 0.0033 from ff chart (Perry)*

$$\Delta P/100\,ft = \frac{2flu^2\rho_L}{g_c D_i (144)} = \frac{2(0.0033)(8.9)^2(46.4)}{32.2(0.5052)(144)} = 1.04\ psi\ per\ 100\,ft$$

Assume Dia = 4 in std = 0.3355 ft, A_C = 0.088 ft², ε/D_i = 0.00015/0.3355 ≐ 0.00045

$u = \frac{1.783}{0.0884} = 20.17$ ft/s which is too high. ∴ 4 in pipe is not suitable.

8 in id. pipe will give ΔP < 1.04 si calculated for 6 in pipe.

6 in std. pipe Schedule 40 is to be recommended.

Answer is (B)

THERMODYNAMICS

Combustion:

B2:

First calculate heats of reaction for each reaction from the given heats of formation using the relation

$$\Delta H_R = \Sigma(\Delta H_f)_P - \Sigma(\Delta H_f)_R$$ Heats of reactions
kcal/gmol @25 ºC, 1 atm

(1) $CH_3CHO(g) + H_2(g) \rightarrow C_2H_5OH(g)$ $\Delta H_1 = -12.51$

(2) $C_2H_5OH(g) + 3O_2(g) \rightarrow 2CO_2(g) + 3H_2O(l)$ $\Delta H_2 = -340.83$

(3) $H_2(g) + \frac{1}{2}O_2(g) \rightarrow H_2O(l)$ $\Delta H_3 = -68.32$

(4) $\quad H_2O(l) \to H_2O(g) \qquad\qquad \Delta H_4 = +10.52$

From first 3 reactions, by adding first 2 and subtracting 3, get the following

(1) $\quad CH_3CHO(g) + H_2(g) \to C_2H_5OH(g) \qquad \Delta H_1 = -12.51$
(2) $\quad C_2H_5OH(g) + 3O_2(g) \to 2CO_2(g) + 3H_2O(l) \quad \Delta H_2 = -340.83$
(3) $\quad -[\ H_2(g) + \tfrac{1}{2}O_2(g) \to H_2O(l)] \qquad\quad -[\Delta H_3 = -68.32]$

(5) = (1) + (2) - (3)

(5) $\quad CH_3CHO(g) + 2\tfrac{1}{2}O_2 \to 2CO_2(g) + 2H_2O(l)$

$\Delta H_c = \Delta H_1 + \Delta H_2 - \Delta H_3$

$\qquad = -12.51 - 340.83 + 68.32 = -285.02 \qquad$ kcal/gmol

Answer is (A)

B3:

Net heating value :

$CH_3CHO(g) + 2\tfrac{1}{2}O_2 \to 2CO_2(g) + 2H_2O(l) \qquad \Delta H_5 = -285.02$
$2\ H_2O(l) \to 2H_2O(g) \qquad\qquad\qquad\qquad\qquad 2\ \Delta H_4 = 2(10.\ 2)$

$CH_3CHO(g) + 2\tfrac{1}{2}O_2 \to 2CO_2(g\ + 2H_2O(g) \qquad \Delta H = -263.98$

Vapor pressure of water at 60 °F and atm. pressure = 13.3 mm Hg
Partial pressure of gas = 760 - 13.3 = 746.7 mm Hg
Molal volume at 60 °F and 746.7 mm Hg = $359 \times \frac{760}{746.7} \times \frac{520}{492}$

$= 386.2$ ft³/lb mol

Net heating value = $\frac{263.98 \times 1000 \times 1.8}{386.2} = 1230.4$ Btu/ft³ of gas

Answer is (B)

Economics:

B4:

Total capital investment = Fixed capital + working capital

Therefore,   3000000 = FC + 0.2(3000000)

FC = 3000000 - 0.2(3000000) = 2,400,000 dollars

Annual sales = (1) (2400000) = 2,400,000 dollars

Annual product cost = $ 180,000

Gross earnings = 2400000 - 180000 = $600000

Gross earnings as % of total capital investment = (600000/3000000)(100) = **20 %**

Answer is (A)

B5:

Let the final payment be   X

First plan	New plan
$6000 on 12/31/1995	$8000 on 12/31/1998
$4000 on 12/31/1997	$ X on 12/31/1999

The two groups of money values are equivalent.

Taking future worthies of the different payments to a common date,

8000(F/P, 10%, 1) + X = 6000(F/P, 10%, 4) + 4000((F/P, 10%, 2)

8000(1.1) + X = 6000(1.4641) + 4000(1.21)

Solving for X gives, X = **$4824.6**

Answer is (A)

Equilibrium data:

B6:

(Note: Problems A41 and A42 are to be solved together)

Using Van Laar constants, first calculate the liquid phase activity coefficients.

$$\log \gamma_A = \frac{A x_2^2}{\left[\frac{A}{B} x_1 + x_2\right]^2} = \frac{0.6596 \times (0.4)^2}{\left[\frac{0.6596}{0.7371}(0.6) + 0.4\right]^2} = 0.09264 \qquad \gamma_A = 1.2378$$

$$\log \gamma_B = \frac{Bx_1^2}{[x_1 + \frac{B}{A}x_2^2]^2} = \frac{0.7371(0.6)^2}{[0.6 + \frac{0.7371}{0.6596}(0.4)]^2} = 0.2421 \qquad \gamma_B = 1.7461$$

$y_A P = x_A \gamma_A P_1$ Therefore, $y_1 = \frac{x_A \gamma_A P_1}{P} = \frac{0.6\,(1.2378)\,(P_1)}{760} = 0.0009772 P_1$

Likewise, $y_2 = \frac{x_B \gamma_B P_2}{P} = \frac{0.4(1.7461)(P_2)}{760} = 0.000919 P_2$

Also, $y_1 + y_2 = 1$

Since t is not known, the solution is by trial and error.

Trials:

Temp. °C	P₁ mm Hg	P₂ mm Hg	y₁	y₂	y₁ + y₂
80	813	293.4	0.7395	0.2696	1.0631
78	751.3	274.4	0.7342	0.2522	0.9864

By interpolation, $t = \frac{1 - 0.9864}{1.0631 - 0.9864} \times 2 + 78 = 78.36$ °C

This is the boiling point of the mixture at 1 atm. pressure

Answer is (C)

B7: At the temperature calculated in **B6**,

$P_1 = 762.1$ mm Hg and $P_2 = 277.9$ mm Hg

$y_1 = 0.7447$ $y_2 = 0.2554$ $y_1 + y_2 = 1.0001$

Therefore, $y_1 = 0.7447 \div 0.745$

Answer is (A)

Estimation of properties:

B8:

$$h = \frac{0.0867 R T_c}{P_c V} = \frac{0.0867\,(0.732)\,(672)}{42 \times 5} = 0.2012$$

$$T_r = \frac{460 + 212}{369.8 \times 1.8} = 1.012$$

$$Z = \frac{1}{1 - h} - \frac{4.934}{T_r^{1.5}} \frac{h}{1 + h} = \frac{1}{1 - 0.2012} - \frac{4.934}{1.012^{1.5}} \frac{0.2012}{1 - 0.2012} = 0.44$$

Moles of propane = 50/44 = 1.1364 lb mol

Then, $P = \frac{ZnRT}{V} = \frac{0.44(1.1364)(0.732)(672)}{5} = 49.19$ atm = 723.1 psia = **708.4 psig**

Answer is (B)

B9:

Critical constants of ethylene are $t_c = 9.7\ °C$, $P_c = 50.9$ atm.

PV = ZnRT

$T_c = 273 + 9.7 = 282.7$ K $= 282.7(1.8) = 508.9\ °R$
$P_c = 500$ atm. $P_r = \frac{500}{50.9} = 9.823$ $T_r = \frac{383.1}{282.7} = 1.36$

From high pressure Z chart, $Z \cong 1.15$

$n = \frac{PV}{ZRT} = \frac{500(2)}{1.15(0.732)(508.9)} = 2.3343$ lb mol

$= 2.3343(28) =$ **65.4 lb**

Answer is (C)

Estimation of thermodynamic properties:

B10:

Normal boiling point of benzyl acetate from Antoine equation

$\ln 760 = 16.5956 - \frac{4104.84}{T - 74.56}$

From which $T_B = 486.6$ K
$T_{br} = \frac{486.6}{699} = 0.6961$ $P_r = \frac{1}{31.4} = 0.032$

From Z chart, $Z \doteq 0.97$

Substituting the appropriate values in the reduced equation,

$\frac{\Delta H_V}{ZRT_C} = \frac{B}{T_C}\left[\frac{T_r}{T_r + C/T_C}\right]^2 = \Delta H_V = ZRB\left[\frac{T_r}{T_r + C/T_C}\right]^2$

$= (0.97)(8.314)(4104.84)\left[\frac{0.6961}{0.6961 + \left(\frac{-74.56}{699}\right)}\right]^2$

$= 46170$ J/g mol

$= 19853$ Btu/lb mol = **132.2 Btu/lb**

Answer is (B)

B11:

$$\bar{H} = 8560 \times 1.8 = 15410 \text{ Btu/lb mol}$$

By definition $H = U + PV$, Therefore, $U = H - PV$

To calculate specific volume at 1000 psia and 400 °F.

$$T_r = \frac{400 + 460}{420 \times 1.8} = 1.14 \qquad P_r = \frac{1000}{39.7 \times 14.7} = 1.71$$

Using Pitzer's correlation, $Z^0 = 0.51$, $Z^1 = 0.13$

Then $Z = Z^0 + \omega Z^1 = 0.51 + 0.187(0.13) = 0.53$

$$V = \frac{0.53(10.73)(860)}{1000} = 4.89 \text{ ft}^3/\text{lb mol}$$

$$U = H - PV = 15410 - \frac{1000(4.89)(144)}{778} = 14505 \text{ Btu/lb mol} = \mathbf{96.6 \text{ Btu/lb}}$$

Answer is (C)

Thermodynamic charts and data bases:

B12:

At beginning. $P_1 V = Z_1 n_1 RT$ \qquad After withdrawal, $P_2 = Z_2 n_2 RT$

$$\frac{P_2 V}{P_1 V} = \frac{Z_2 n_2 RT}{Z_1 n_1 RT}$$

Since V and T are constant, the above reduces to $\frac{P_2}{P_1} = \frac{Z_2 n_2}{Z_1 n_1}$

Now $\frac{n_2}{n_1} = \frac{5000/MW}{15000/MW} = \frac{1}{3}$ \qquad MW = molecular weight of ethylene

Thus $P_2 = \frac{1}{3}\frac{Z_2}{Z_1} P_1$

Now $P_{r1} = \frac{1064.7}{50.9(14.7)} = 1.423 \qquad T_{r1} = T_{r2} = \frac{273 + 24}{282.7} = 1.0506$

Reading from Z chart, we get $Z_1 = 0.35$

With the above Z value, $P_2 = \frac{1064.7}{3(0.35)} Z_2 = 1014 Z_2$

To find P_2, a trial and error solution is required.

Try $P_2 = 650$ psia. $P_{r2} = 650/(50.9 \times 14.7) = 0.0.87$ Z_2 from chart = 0.65

$P_2 = 0.65(1014) = 659.1$ psia

Try $P_2 = 659.1$ psia. $P_{r2} = 659.1/(50.9 \times 14.7) = 0.65$ Z_2 from chart = 0.65

$P_2 = 0.65(1014) = 659.1$ psia

Pressure = 659.1 - 14.7 = 644.4 psig say 645 psig

Answer is (D)

B13:

On Mollier diagram, locate the point 200 psia and 400 °F temperature.
Read H = 1210.2 Btu/lb and S = 1.56 Btu/(lb°R)
Since process is adiabatic and reversible, $\Delta S = 0$.
Therefore S = 1.56 Btu/(lb°F)
To find % moisture, on Mollier diagram, move along vertical at S = 1.56
and locate intersection with $P_2 = 14.7$ psia line.
Approximate % moisture = 11.3 %
Therefore vapor quality of steam = 100 - 11.3 = 88.7 %

Answer is (B)

V-L Composition of miscible and partially miscible systems:

B14:

Since the two liquids are insoluble, they exert their own vapor pressures in the vapor phase. By inspection of the given vapor pressure data, it is clear that the boiling point is closer to the boiling point of water since the vapor pressures of nitrobenzene are very small. Solution of finding BP is trial and error procedure.

Assumed temperature, °C	vapor pressure of nitrobenzene mm Hg	vapor pressure of water mm Hg	total pressure mm Hg
99.0	20.1	733.2	753.3
99.3	20.4	741.1	761.5
99.2	20.3	738.5	758.8
99.25	20.3	739.8	760.1 close

Therefore, BP is 99.25 °C

Answer is (A)

B15:

$$\text{Mol fraction of nitrobenzene in vapor} = \frac{\text{vapor pressure of nitrobenzene at BP}}{\text{Total pressure}}$$

$$= \frac{20.4}{760} = 0.02676$$

Answer is (B)

Thermodynamic laws:

B16:

2 hp = 2 (2545) = 5090 Btu

By heat balance, $W = Q_h - Q_L$ therefore $Q_h = 5090 + 7500 = 12590$ Btu/h

By second law of thermodynamics, $\frac{Q_h}{Q_L} = \frac{T_h}{T_L} = \frac{T_h}{520} = \frac{12590}{7500} = 1.6787$

Therefore, $T_h = 1.6787(520) = 873\ ^\circ R = \mathbf{413\ ^\circ F}$

Answer is (C)

END OF SOLUTIONS OF PM BONUS EXAM..